T0271864

Design of Green Liquid Dielectrics for Transformers: An Experimental Approach

Biodegradable Insulating Materials for Transformers

RIVER PUBLISHERS SERIES IN BIOTECHNOLOGY AND MEDICAL RESEARCH

Series Editors:

PAOLO DI NARDO
University of Rome Tor Vergata
Italy

PRANELA RAMESHWAR
Rutgers University
USA

Aiming primarily at providing detailed snapshots of critical issues in biotechnology and medicine that are reaching a tipping point in financial investment or industrial deployment, the scope of the series encompasses various specialty areas including pharmaceutical sciences and healthcare, industrial biotechnology, and biomaterials. Areas of primary interest comprise immunology, virology, microbiology, molecular biology, stem cells, hematopoiesis, oncology, regenerative medicine, biologics, polymer science, formulation and drug delivery, renewable chemicals, manufacturing, and biorefineries.

Each volume presents comprehensive review and opinion articles covering all fundamental aspect of the focus topic. The editors/authors of each volume are experts in their respective fields and publications are peer-reviewed.

For a list of other books in this series, visit www.riverpublishers.com

Design of Green Liquid Dielectrics for Transformers: An Experimental Approach

Biodegradable Insulating Materials for Transformers

T. Mariprasath

K.S.R.M College of Engineering (Autonomous), India

Victor Kirubakaran

The Gandhigram Rural Institute-Deemed to be University
Tamil Nadu, India

Perumal Saraswathi

A.P.C Mahalaxmi College for Women, India

Kumar Reddy Cheepati

K.S.R.M College of Engineering (Autonomous), India

Prakasha Kunkanadu Rajappa

Catalonia Institute for Energy Research IREC, Barcelona, Spain

Published 2024 by River Publishers
River Publishers
Alsbjergvej 10, 9260 Gistrup, Denmark
www.riverpublishers.com

Distributed exclusively by Routledge
605 Third Avenue, New York, NY 10017, USA
4 Park Square, Milton Park, Abingdon, Oxon OX14 4RN

Design of Green Liquid Dielectrics for Transformers: An Experimental Approach / by T. Mariprasath, Victor Kirubakaran, Perumal Saraswathi, Kumar Reddy Cheepati, Prakasha Kunkanadu Rajappa.

Routledge is an imprint of the Taylor & Francis Group, an informa business

ISBN 978-87-7004-152-2 (hardback)
ISBN 978-87-7004-204-8 (paperback)
ISBN 978-87-7004-215-4 (online)
ISBN 978-87-7004-208-6 (master ebook)

Contents

Preface

In an era where sustainable practices are imperative for the preservation of our planet, the realm of electrical engineering faces a critical challenge: how to ensure reliable and efficient energy transmission without compromising our environment. The transformative shift toward green insulating materials for transformers is a pivotal step in this direction. It is within this context that the present volume, "Green Insulating Materials for Transformer: An Experimental Approach," emerges as a comprehensive guide for researchers, engineers, and enthusiasts invested in sustainable energy solutions.

This book encapsulates a thorough exploration of the realm of eco-friendly liquid dielectrics, shedding light on their importance, historical development, and crucial characterization techniques. Delving deep into the intricate web of material analysis, it traverses the realm of spectroscopy, thermal aging analysis, morphological studies, crystalline structures, and the implications of thermal stress and contamination on liquid dielectrics. The exploration of various condition monitoring techniques, with a special focus on the application of thermal imagery, further underscores the significance of real-time assessment in ensuring the longevity and efficiency of transformers.

Moreover, in an age where technology evolves at an unprecedented pace, the integration of artificial intelligence and machine learning algorithms to predict the lifetime of transformers stands as a testament to the power of innovation in bolstering sustainable infrastructure. Chapter by chapter, this book not only elucidates the significance of environmentally conscious practices in the field of electrical engineering but also paves the way for a future where green solutions are not just a choice but an imperative necessity as well.

The proposed book comprises seven chapters. Chapter 1 discusses the importance of transformers and their insulation systems, emphasizing the necessity of alternative liquid dielectrics.

In Chapter 2, we conduct an extensive literature review drawing from reputable journals. In Chapter 3, we provide a historical review of the importance of spectroscopy analysis. Chapter 4 delves into the preparation of materials for testing, followed by an experimental analysis of critical characteristic assessment methods. It also explores the role of spectroscopy analysis in evaluating the degradation rate of cellulosic insulating materials used in transformers. This chapter provides valuable information on how to write inferences for SEM, XRD, and FTIR analyses on test samples.

Chapter 5 offers a critical review of the significance of monitoring the condition of transformers, and Chapter 6 introduces an efficient condition monitoring approach using thermal imagers and the hotspot indication method.

The final chapter, Chapter 7, sheds light on the pivotal role of artificial intelligence in assessing transformer insulation systems. It also delves into the application of machine learning algorithms for estimating the life cycle of transformers.

Throughout the book, you will find in-depth information on recent challenges in designing transformer insulation systems, guidance on evaluating critical properties of eco-friendly insulating materials, insights into efficient condition monitoring methods, and methods for predicting the life cycle of transformers using machine learning. The book provides readers with practical knowledge on transformer insulation system design, effective monitoring techniques, and the role of machine learning in assessing transformer condition and estimating their life cycle.

List of Figures

List of Tables

List of Abbreviations

AI	Artificial intelligence
BDV	Breakdown voltage
BET	Brunauer-Emmett-Teller
CHTC	Convective Heat Transfer Coefficient
C-index	Concordance index
CSO	Cottonseed oil
DCT	Dielectric constant
DC	Direct current
DVC	Dynamic viscosity coefficient
DDB	dodecylbenzene oil
DDF	Differential distribution factor
DGA	Dissolved gas analysis
DT	Distribution transformer
EDL	Electric Double Layer
E-TRM	Electro-thermal resistance model
FTIR	Fourier transform infrared spectroscopy
h-BN	Hexagonal boron nitride
HMWA	Higher molecular weight fatty acids
HPLC	High-Performance Liquid Chromatography
IEC	International electrotechnical commission
IS	International standard
KMS	Kalahari melon seed
LCA	Life cycle assessment
LCI	Loading capacity increment
LMA	Lower Molecular Acids
LMWA	Low molecular weight acids
LMWA	Low molecular weight antioxidants
MAE	Mean absolute error
ML	Machine learning
MO	Mineral oil
MO	Mustard oil

MOWC	Mineral Oil with Catalytic
MSE	Mean squared error
MWCNT	Multi-walled carbon nanotube
NEO	Natural Ester Oil
NEOWC	Natural Ester Oil with Catalytic
NF	Nanofluid
OLTC	On-Load Tap Changer
PD	Partial discharge
PFAE	Perfluoroalkylethyl
ppm	Parts per million
SEM	Scanning electron micrograph
SFRA	Sweep Frequency Response Analysis
SO	Sunflower oil
TAN	Total acid number
TCG	Total Combustible Gas
TEM	Transmission Electron Microscope
TO	Transformer oil
UHF	Ultrahigh-frequency
UHV	Ultra-high voltage
UV	Ultraviolet
VGO	Vegetable Oil
WC	Water Content
XO	Ximenia oil
XRD	X-ray diffraction

1

Introduction

1.1 Necessity of Alternate Liquid Dielectric

Transformers are a vital part of electrical distribution systems, guaranteeing the safe and effective flow of electricity from generation facilities to consumers' homes. Electrical energy can be transmitted across great distances with minimal loss if transformers are used to increase or decrease the voltage. Providing electricity to residences, factories, and commercial buildings would be impractical without power transformers. Equally important are distribution transformers, which convert power from high-voltage transmission lines to the low-voltage networks that supply homes and businesses. They bring the voltage down to acceptable levels so that power can be transmitted to homes and businesses without risk. Maintaining consistent voltage and protecting electrical equipment from overvoltage are two of the most important functions of distribution transformers. To guarantee reliable and secure operation, transformers rely heavily on their insulating systems. To insulate the transformer's windings and other parts, a common insulation scheme utilizes cellulose paper, oil, and pressboard. To ensure the transformer's durability and dependability, these components are meticulously designed and placed to withstand the electrical stresses and temperature changes encountered in operation [1]-[3].

Due to its high dielectric strength and high thermal conductivity, mineral oil has long been used for transformer insulation and cooling. Mineral oils are commonly classified depending on their chemical makeup. Naphthenic mineral oil is recognized for its advantageous characteristics at low temperatures, rendering it a desirable choice for transformers deployed in regions with cold climates. In contrast, paraffinic mineral oil demonstrates exceptional performance in terms of maintaining stability at high temperatures, making it well-suited for transformers that function in environments with elevated temperature levels. Aromatic mineral oil is highly regarded for its exceptional electrical insulating characteristics, particularly in situations involving high voltage. In addition, specialized formulations such as low-pour-point oil and high-flash-point oil are designed to address specific environmental difficulties and safety considerations. The selection of mineral oil variety is contingent upon various aspects, such as the geographical placement of the transformer, its voltage classification, and the anticipated temperature spectrum. The selection of an appropriate mineral oil is crucial in order to achieve optimal performance and prolong the lifespan of transformers, while also maintaining compliance with safety rules and standards. Regular maintenance and oil testing play a vital role in monitoring the oil's condition and the general health of the transformer, hence guaranteeing dependable and effective operation [4], [5].

However, there are major restrictions. In the event of a transformer malfunction or leak, mineral oil can be released into the environment and cause serious damage. It is not biodegradable; therefore, improper disposal could lead to environmental contamination. These restrictions have prompted the exploration of greener substitutes. To lessen the negative effects on the environment caused by mineral oil, biodegradable liquid dielectrics are required in transformers. Natural esters, which are obtained from renewable resources like vegetable oils, are one example of a biodegradable liquid that provides a safer and more sustainable alternative. They do not harm humans or animals, decompose quickly, and do not release dangerous chemicals into the environment like mineral oil does. Switching to biodegradable dielectrics is a crucial step toward lowering transformers' environmental impact and creating a more sustainable energy system overall. In conclusion, power transformers and distribution transformers play a crucial role in the reliable transmission and distribution of electricity across the electrical grid. Their insulating solutions are meticulously crafted to provide risk-free performance. Biodegradable liquid dielectrics provide a more sustainable and environmentally friendly alternative to mineral oil, which has been widely utilized in transformers despite its limits due to its flammability and environmental impact. Adopting these advances is crucial for reducing negative impacts on the environment and creating a more sustainable energy system. Synthetic ester oil, commonly known as ester oil, is a synthetic lubricating oil that finds extensive usage across many industries. The production of this substance occurs via the esterification process, when an organic acid undergoes a reaction with an alcohol, resulting in the formation of ester molecules. The resultant oil possesses various distinct characteristics that render it suitable for particular applications. One notable attribute of synthetic ester oil is its exceptional heat stability and elevated flash point. This characteristic renders it appropriate for utilization in situations characterized by severe temperatures, including both exceedingly low and highly elevated heat levels. The use of this material is frequently observed in several industries, including aviation, where it is employed in jet engines operating under elevated temperatures, as well as in automotive applications, particularly in high-performance and racing engines [6]-[8].

Synthetic ester oils are known to possess exceptional lubricating qualities. The inherent lubricating properties of these substances contribute to the mitigation of friction and wear in machinery and engines, hence resulting in prolonged equipment lifespan and enhanced overall operational effectiveness. In the domain of aviation, lubricants are employed to provide lubrication to essential engine components. In the field of car racing, the use of these components serves the purpose of diminishing engine friction and augmenting horsepower. Another noteworthy characteristic of synthetic ester oil pertains to its chemical compatibility with diverse materials. There is a reduced probability of degradation occurring in rubber seals, gaskets, and other commonly employed materials in machinery. This characteristic renders it appropriate for utilization in hydraulic systems, compressors, and other machinery where the preservation of seal integrity is of utmost importance. Moreover, synthetic ester oil is recognized for its capacity to degrade naturally and its little ecological footprint. The lubricant in question undergoes natural degradation within the environment, rendering it a more environmentally sustainable alternative in comparison to some other lubricants. Due to its environmentally conscious attributes, this substance has garnered significant attention in sectors that prioritize ecological responsibility, such as the utilization of biodegradable hydraulic fluids and environmentally sustainable lubricants [9]-[11].

2

Study on Alternate Oil Properties

2.1 Study on Alternate Oil Properties

Sifeddine Abdi et al. conducted an accelerated thermal aging test on transformer oil, subjecting it to temperatures of 80, 100, 120, and 140 °C for a duration of 5000 hours. The breakdown strength of oil samples was evaluated in accordance with the IEC 60156 standard to evaluate the electrical properties of transformer oil at increased temperatures, with a sampling interval of 500 hours. The initial breakdown strength of oil was measured to be 80 kV. However, this value was found to decrease by 50% within the temperature range of 80–100 °C during the aging process. Furthermore, when the oil was subjected to thermal aging at 120 °C, the breakdown strength was seen to decrease by 80%. The temperature increase resulted in a significant decrease in the breakdown voltage (BDV) of the transformer oil, reaching a critical threshold after a sample time of 3000 hours. The development of water molecules due to oil warming stimulates the production of gas bubbles inside the oil. The presence of these contaminants gives rise to a phenomenon known as streamer formation within the gas bubbles. Propagation occurs when electricity is applied. The aforementioned procedure undergoes expansion, resulting in subsequent disintegration. The electrical characteristics of transformer oil at high temperatures were determined by measuring the DDF of oil samples at 500-hour intervals using an Automatic Dissipation Factor and Resistivity Test Equipment Dieltest DTL system, as per IEC 62247. The experimental results show that the DDF of the oil sample is 0.0025 at temperatures below 100 °C. The DDF of the oil sample is 0.0165 at temperatures over 100 °C, which is much higher than the recorded value at 80 and 100 °C but still within an acceptable range. An increase in ionic mobility due to heating causes a rise in conduction loss. The oxidation reactions within the oil are decreasing the oil's viscosity. During the entirety of the aging process at lower temperature ranges, specifically at 80, 100, and

120 °C, the acidity level of transformer oil experiences an increase. However, it is important to note that this increase remains within acceptable limits, as the stability of the oil remains unaffected. In instances where the acidity of oil surpasses the permissible thresholds for sample intervals lasting 2000 hours at a temperature of 140 °C, the recorded acidity of the oil is 1.2 mg Cu/g. The thermal degradation of insulating oil can be inferred due to the promotion of oxidation at higher temperatures, resulting in the development of acid inside the oil. The water content of an oil sample was determined using the automatic Karl Fischer titration method in accordance with the IEC 60814 standard, with a sampling interval of 500 hours. The water content of oil remains within acceptable levels throughout the entire aging period when sampled at temperatures of 80 and 100 °C. However, at higher temperatures ranging from 120 to 140 °C, the water content experiences a significant increase. This increase reaches a critical value after 3000 hours of sampling, primarily due to the decreased stability of the oil. Consequently, the oil decomposes and oxidizes upon overheating at elevated temperatures. The initial measurement of oil viscosity is recorded as 6.998 centistokes at a temperature of 40 °C. Nevertheless, it should be noted that the viscosity of the oil sample is subject to variation depending on the sampling temperature and time interval. The decrease rate exhibits a significant increase when subjected to an aging temperature of 80 °C for a sampling duration of 2500 hours. In contrast, it was observed that the viscosity of the oil sample significantly increased to an elevated level when exposed to a temperature of 120 °C for a duration of 5000 hours, as reported in reference [12].

In a study conducted by Al-Eshaikh et al., the electrical properties of maize oil were analyzed, and its potential as a liquid dielectric was assessed. The measurement of the breakdown strength of the oil samples is conducted in accordance with the IEC 60156 standard, employing the Foster type 60A automatic oil tester. The average breakdown voltage (BDV) of mineral oil is 33.9 kV, with a standard deviation of 3.96. In contrast, the BDV of maize oil is 46.1 kV, with a standard deviation of 4.7 kV. The mineral oil sample exhibits a moisture level of 15 parts per million (ppm), while the moisture content of maize oil is measured at 127 ppm. The available evidence suggests that the moisture content does not have a significant impact on the breakdown strength of maize oil. This conclusion is drawn from the fact that maize oil contains over 70% unsaturated fatty acids, as reported in the reference Tettex precision Schering bridge system used to quantify maize oil by M. A. Al-Eshaikh et al. Corn oil has a far higher neutralization number than mineral oil does at 80 °C because its DDF is nearly double that of mineral oil at ambient temperature. Corn oil DDF grows dramatically from 0.2% to 2.7%, much surpassing mineral oil DDF. This is because the insulating substance is so sturdy. The DDF of the sample aged at 110 °C for 600 hours with catalytic addition has also been tested. Since maize oil contains glycerol and corporeal groups, its DDF is initially higher than that of mineral oil. Since hydrocarbons make up the bulk of mineral oil, DDF quickly climbs as temperature does after 300 hours of sampling. According to this evidence, maize oil is a more reliable alternative than mineral oil. The measurements were performed using an Oswald viscometer with a precision of 0.1 in accordance with the ASTM D-455 standard. In this experiment, the viscosity of the oil sample was determined by calculating the ratio of its dynamic viscosity to its density. Kinematic viscosity pertains to the ability of a particular oil to flow at a specified temperature. At ambient temperature, the viscosity of maize oil is comparatively greater than that of mineral oil, resulting in a reduced cooling capacity for the former. In contrast, the oxidation of minerals within the transformer results in the formation of sludge, hence diminishing the oil's heat transfer capacity in the transformer. In contrast, it has been observed that vegetable oil does not undergo sludge formation during the process of oxidation. The measurements were performed using an Oswald viscometer with a precision of 0.1 in accordance with the ASTM D-455 standard. In this study, the viscosity of the oil sample was determined by calculating the ratio of its dynamic viscosity to its density. Kinematic viscosity pertains to the ability of a particular oil to flow at a specified temperature. At ambient temperature, the viscosity of maize oil is greater than that of mineral oil, resulting in a reduced cooling capacity for the former. In contrast, the oxidation of minerals within the transformer results in the formation of sludge, thereby diminishing the oil's capacity to dissipate heat. In contrast, vegetable oil does not undergo sludge formation during the process of oxidation [13].

In their study, Liao et al. performed an accelerated thermal aging test on BIOTEMP oil, which had catalytic additives. The test was conducted at a temperature of 170 °C for a duration of 216 hours. The breakdown voltage (BDV) of the oil samples was measured according to the ASTM D1816 standard. The results indicated that the BDV of the oil samples was influenced by the concentration of moisture and furfural at the initial stage and by the acid content at the later stage. Natural ester oil has a higher acid content in comparison to mineral oil. However, this disparity does not have an impact on the breakdown strength of natural ester oil. This is due to the generation of mild long-chain fatty acids during the thermal aging process of natural ester oils, which do not possess corrosive properties. In addition, it produces peroxide, a compound that has a strong attraction to hydrogen gas and effectively mitigates the production of bubbles inside the oil. Oil's DDF needs to be gradually raised over the first 150 hours of aging. The increase in charged particles per unit of oil volume also led to a quick rise in the DDF of the oil sample. There is an increase in moisture and furfural content at the outset, followed by a marked decrease as the wine ages. Therefore, DDF of oil relies on acids with a larger molecular weight after the initial 150 hours of sampling. An experimental thermal aging experiment on vegetable oil using NOMEX paper and pressboard, objective of this experiment was to analyze the acid contents of the oil samples. The experiment was conducted at a temperature of 170 °C for a total duration of 216 hours, during which sampling was carried out periodically. The concentration of acids in vegetable oil demonstrates an upward trend as the aging period progresses. Specifically, after 192 hours of sampling, there is a notable exponential increase in acid content. This can be attributed to the production of acids with greater molecular weights by vegetable oil. The solubility of the substance is higher in oil compared to paper. As a result, the interaction between the substance and the cellulose insulating material is limited, leading to a low rate of degradation of the paper material. The measurement of viscosity for each oil sample must be conducted at regular intervals of 50 hours. The oil sample's viscosity at a temperature of 80 °C falls within the specified range of 11−13 cSt, meeting the standard limit. However, it should be noted that the viscosity of oil exhibits a significant increase when subjected to a temperature of 40 °C in the presence of circulating dry air. The oxidation of solid and liquid insulation over time leads to an elevation in furfural concentrations and acidity within the oil. In addition, polymerization occurs in order to increase the viscosity of oil. Therefore, in low-voltage applications, the thermal conductivity of oil is diminished. However, it should be noted that the viscosity of oil exhibits a significant increase when subjected to a temperature of 40 °C in the presence of circulating dry air. The oxidation of solid and liquid insulation over time leads to a rise in furfural concentration and acidity in the oil. In addition, polymerizations occur to increase the viscosity of oil. Therefore, it can be observed that the heat transmission capability of oil is diminished in low-voltage applications. An accelerated thermal aging test was performed on a sample of vegetable oil with the addition of a catalytic substance (NOMEX) at a temperature of 170 °C for a duration of 216 hours. The purpose of this test was to measure the viscosity of the oil sample according to the ASTM D445 standard. The measurement of oil sample viscosity is required at regular intervals of 50 hours during the sampling process. The oil sample's viscosity at a temperature of 80 °C falls within the specified range of 11–13 cSt, meeting the standard limit. However, the viscosity of oil exhibits a significantly greater magnitude at a temperature of 40 °C when subjected to the influence of flowing dry air. The oxidation of solid and liquid insulation leads to a progressive increase in furfural concentration and acidity in the oil. In addition, polymerizations occur to increase the viscosity of oil. Therefore, in low-voltage applications, the thermal conductivity of oil is diminished. However, it is observed that the viscosity of oil exhibits a significant increase when subjected to a temperature of 40 °C in the presence of circulating dry air. The oxidation of solid and liquid insulation over time leads to a rise in furfural concentration and acidity in the oil. In addition, polymerization occurs in order to increase the viscosity of oil. Therefore, it can be observed that the heat transmission capacity of oil is diminished in low-voltage applications [14].

Singh et al. conducted a study to examine the breakdown voltage (BDV) of 10 different transformer oils in order to assess the impact of aging on the electrical properties of the oils. The breakdown voltage (BDV) of the oil sample was determined in accordance with the specifications outlined in IS 335. The measurement was conducted using a 12.5-mm sphere electrode with a gap separation of 2.5 mm. Based on the findings obtained from the conducted experiments, it can be observed that the breakdown voltage (BDV) of transformer oil exhibits a gradual drop over time, displaying a non-linear correlation with the aging process. During the operational state of a transformer, electrochemical stress arises within the transformer, leading to the degradation of its insulating. These mechanisms efficiently augment the quantity of conducting particles, which undergo rapid augmentation as a result of the aging process. Therefore, the presence of contaminants significantly influences the BDV (breakdown voltage) of oil samples [15].

Abderrazzaq et al. tested olive oil breakdown strength using IEC 60156. The test results were compared to mineral oil. The breakdown strength was measured at 25 °C using a 2.5-mm sphere gap electrode. The average breakdown strength of residual mineral oil is 59.2 kV, which surpasses the ASTM D877 norm by 13.8%. Mineral oil reduces breakdown strength to 24 kV, which is 49% of the IEEE Std C57.106-2002 limit. Transformer mineral oil is contaminated by electromechanical stress. The breakdown strength of unfiltered fresh olive oil is 14.5 kV, while aged oil is 10.9 kV. The electrical breakdown strength of freshly filtered olive oil is 36 kV, while that of old oil is 19 kV. Multi-filtration is used on natural, irrigated olive oils. This technique increases the mean breakdown strength of naturally grown olive oil to 37.5

kV. In contrast, irrigated olive oil breaks down at 57.75 kV. This makes it ideal for power transformers. According to reference, filtration removes carbon, free water, sludge, suspended particles, and acidity from oil. Freshly filtered (aged) olive oil has 446 (520) ppm water content, while unfiltered has 861 (918) ppm. Olive oil from natural growth and irrigation has 1127/1796 ppm water content. Transformer oil water content cannot exceed 50 ppm. The goal is achieved through reclamation. Heating and filtering reduce olive oil moisture. Heating significantly reduces olive oil's water content, whereas cooling increases it. New olive oil is filtered to reduce moisture content by 93%, but old oil is reduced by 76%. According to prior research, filtering is an effective method for removing water from oil. New and aged olive oils have 9.9 and 11.6 cSt viscosity. Additionally, tree oil from organically grown trees has a viscosity of 8.43 cSt, whereas oil from irrigated olive trees has 8.5 cSt. Increased oil temperature lowers viscosity and raises transformer pressure. Thus, temperature and pressure affect oil viscosity. Fresh and aged olive oils have 9.9 and 11.6 cSt viscosity. Additionally, tree oil from organically grown trees has a viscosity of 8.43 cSt, whereas oil from irrigated olive trees has 8.5 cSt. Increased oil temperature lowers viscosity and raises transformer pressure. Thus, temperature and pressure affect oil viscosity [16].

Liu and Wang [17] conducted a study in accordance with the IEC standard, and D. Wang conducted an assessment of the breakdown and withstand strength of three types of insulating fluids: natural ester (FR3), synthetic ester (Midel 7131), and mineral oil (Gemini X). The oil samples were subjected to the application of breakdown voltage using a 10-stage Haefely impulse generator, employing a sphere-to-sphere electrode design. The device is capable of producing a lightning impulse with a duration of 1.2 microseconds and a switching impulse with a duration of 250–2500 microseconds. In the first stage, oil samples were subjected to a negative lightning impulse voltage. The Gemini X oil sample exhibits a greater mean breakdown strength of 243.9 kV, while the Midel 7131 oil sample demonstrates a slightly lower mean breakdown strength of 208.8 kV. The FR3 oil sample, on the other hand, displays the lowest mean breakdown strength among the three, measuring at 202.8 kV [17].

X. Wang conducted a study on. Breakdown strength for mineral oil (Gemini X), synthetic ester (Midel 7131), and natural ester (FR3) was investigated by D. Wang in accordance with the ASTM D1816 standard. The present study aimed to evaluate the impact of cellulosic materials, copper, and water content on breakdown strength. The oil samples were subjected to the application of breakdown voltage. The experimental procedure utilized the

Baur 75 automatic breakdown tester, equipped with a spherical electrode of 36-mm diameter and a gap of 1 mm. Based on the findings obtained from the conducted experiments, it was seen that the mean breakdown strength of the mineral oil sample, after undergoing filtration and dehydration, was measured to be 47.7 kV. In contrast, the breakdown strengths for FR3 and Midel 1713 were determined to be 44.5 and 45.1 kV, respectively. The breakdown strength of the treated samples is greater than that of the unprocessed samples. The pure oil samples are supplemented with cellulosic material. As a result, the particle content of oil is elevated. When the value of a mineral oil sample surpasses 200,000, it is necessary to decrease the breakdown strength to 50% of that observed in clean oil samples. This reduction is lower compared to synthetic ester and FR3 oil. When the quantity of copper particles reached 3000, the breakdown strength of Gemini X oil decreased to 40% of that observed in clean oil. In comparison, the breakdown strength of synthetic ester was 70%, while FR3 exhibited a breakdown strength of 73%. This suggests that FR3 oil has higher breakdown strength compared to mineral oil, even in the presence of contamination from metallic components and cellulosic materials within the transformer. The low viscosity of mineral oil enables the movement of particles more effectively compared to ester oil, as a result of the observed occurrences. The oil exhibits a low relative moisture content, wherein water molecules form weak hydrogen bonds with polar hydrocarbon molecules. On the other hand, it is noteworthy that the mineral exhibits a comparatively elevated level of moisture content. It is seen that certain water molecules, which are released from the oil, assume the role of charge carriers. According to reference [18], the presence of oil samples decreases the breakdown strength.

The breakdown strength test on HONE and mineral oil was conducted by Santanu Singha et al., following the guidelines outlined in ASTM D 1816. In comparison to mineral oil, the dielectric constant (DC) of HONE is consistently higher across all stages of aging. Prior to a 1000-hour aging period, the direct current (DC) of the oil samples exhibits instability, which can be attributed to the presence of by-products resulting from the aging process of cellulose insulation and oil. After a period of aging exceeding 1000 hours, it has been observed that the direct current (DC) of the oil samples remains constant. The measurement was conducted in accordance with the guidelines outlined in ASTM D974. During the initial 500-hour sampling period, it is observed that the acidity content of mineral oil exhibits an upward trend as it ages. This increase can be attributed to the decomposition of Kraft paper, which results in the generation of low molecular weight acids (LMWA) that subsequently mix with the oil samples. Moreover, as the duration of aging increases, the acidity levels of oil samples become saturated. Consequently, the low molecular weight antioxidants (LMWA) are unable to permeate into the oil and instead remain within the Kraft paper. This phenomenon contributes to an accelerated rate of paper aging. The tensile strength of paper experiences a significant decrease following a 500-hour sampling period, resulting in severe degradation. This degradation is assessed using the DuPont TM 0659-98 method as outlined in the ASTM D828 standard. In contrast, the acidity of HONE oil exhibits a small increase over a period of 336 hours. Moreover, there is a significant rise observed throughout the aging process, as the LMA acid is removed from cellulose insulation and subsequently diffuses into the oil samples after 500 hours of sampling intervals. Hence, it can be shown that the rate of decline in tensile strength of paper reaches saturation point after a sample period of 500 hours. The moisture content of mineral oil experiences a minimum rise of 250% after being aged for a duration of 3000 hours. In contrast, the moisture content of HONE undergoes an increase within the range of 336–672 hours of aging. After a duration of 672 hours, the quantity of water molecules in HONE experienced a reduction. This can be attributed to the occurrence of a hydrolysis reaction involving water molecules contained in the oil, leading to the consumption of water molecules. According to a study cited as reference. The researchers also conduct a comparative analysis with high oleic natural ester. During the initial 1500 hours of operation, the viscosity of mineral oil and HONE oil exhibits temporal instability. The viscosity of HONE and mineral oil remained relatively stable during 3500 hours of operation due to the experimental conditions, specifically the use of a nitrogen atmosphere. This controlled environment minimized the impact of oxygen on the ester oil, making high oleic natural ester oil a favorable choice for applications involving hermetically sealed transformers. During the initial 1500 hours of operation, the viscosity of mineral oil and HONE oil exhibits a lack of stability. The viscosity of HONE and mineral oils remained relatively stable during 3500 hours of operation. This can be attributed to the test being conducted under a nitrogen atmosphere, which effectively minimized the impact of oxygen on ester oil. Consequently, high oleic natural ester oil demonstrates significant suitability for applications involving hermetically sealed transformers [19].

Biotemp oil exhibited a higher breakdown strength than mineral oil during the 30-day test. After 80 days, Biotemp's breakdown strength decreased due to the oil's relative moisture and acidity increasing. However, mineral oil breakdown strength is reduced to 20 days during sampling. Mineral oil has a higher contamination rate than natural ester oil. Natural ester oil impregnated pressboards have a lower permittivity ratio than mineral oil impregnated ones. Because of this, natural ester oil with paper impregnated insulating system minimizes oil duct strain, reducing creep discharge. Natural ester oil also protects cellulosic insulating materials, slowing their degradation. Increased temperature promotes peroxide production due to its affinity for hydrogen gas, reducing bubbles. Liao et al. measured the relative permittivity of mineral oil, mineral oil impregnated pressboard, and natural ester oil. Measurements were conducted using Navo Control Concept 80 broadband dielectric spectroscopy from 20 to 90°. The relative permittivity of natural ester oil was higher than mineral oil at all temperatures. Nearby oil-impregnated pressboard has higher permittivity. The oil-impregnated pressboard-to-oil ratio is lower than other materials. The oil channel or oil wedge will have less tension due to this event. It prevents creepage discharge in natural ester-impregnated insulation systems. Measurements were taken using Navo Control Concept 80 broadband dielectric spectroscopy from 20 to 90°. The relative permittivity of natural ester oil was higher than mineral oil at all temperatures. The permittivity of oil-impregnated pressboard is similar. The oil-impregnated pressboard-oil ratio is lower than other materials. The oil channel or oil wedge will have less tension due to this event. Thus, it limits creepage discharge in natural ester-impregnated insulation systems. Comparing mineral oil with ester oil's relative and absolute moisture content across different aging durations, ester oil has a higher absolute moisture content than mineral oil during aging. Mineral oil has significant wetness except for the first 30 days. Ester oil had 1.25% moisture by weight after 58 days. Moisture hydrolyzes ester oil, consuming its dissolved water content [20].

In their study, Jeong et al. (year) conducted an investigation into the breakdown strength of vegetable and mineral oil samples that had undergone aging. Sample 1, consisting of vegetable oil and mineral oil, must undergo an aging duration of about 1674 hours at a temperature of 30 °C. Similarly, sample 2, comprising only of mineral oil, must be subjected to an aging period of 756 hours at the same temperature. During the process of aging, the laboratory employed a suitable experimental arrangement to ensure that the temperature of the highest level of oil remained at 120 °C, while the temperature of the lowest level of oil was kept at 30 °C. The oil sample was subjected to a breakdown voltage test utilizing two copper wires, each with a diameter of 12.5 mm and spaced 2.5-mm apart. During the entire aging process, it was observed that vegetable oil samples exhibited higher levels of acidity and water content compared to mineral oil samples. However, it is worth noting that the breakdown strength of vegetable oil is significantly greater when compared to samples of mineral oil. Due to the composition of carbon and hydrocarbon molecules in vegetable oils, they undergo a reaction with oxygen. However, this reaction results in the formation of a relatively small amount of sludge. In contrast, mineral oil generates a significantly larger quantity of sludge. The presence of sludge in mineral oil samples has been found to accelerate the process of breakdown, leading to quicker degradation [21].

In a study conducted by Saruhashi et al., the authors investigated the breakdown strength of natural ester, synthetic ester, and mineral oil samples that had undergone thermal aging for 1000 hours at temperatures of 130 and 180 °C. The breakdown strength was measured using a 2.5-mm rod electrode. The oil samples were subjected to a negative lightning impulse voltage. The natural ester oil exhibits a comparatively high breakdown strength, which is subsequently followed by silicone and synthetic ester oil. As thermal aging progresses, there is a gradual decrease in the breakdown strength of oil samples. The average breakdown value of fresh natural ester oil is recorded as 172 kV, whereas silicone oil exhibits a mean breakdown value of 153 kV, and synthetic ester oil demonstrates a mean breakdown value of 121 kV. On the contrary, the average breakdown strength of matured natural ester oil is recorded as 138 kV, whereas silicone oil exhibits a slightly higher value of 143 kV. The breakdown strength of aged oil exhibits a decrease of 10% for silicone oil, 20% for synthetic ester oil, and 30% for natural ester oil. This suggests that the reduction in breakdown strength for silicone oil is relatively low. It is worth noting that, during the aging process, the total acid number of natural ester oil was found to be higher compared to the other oils. The samples were subjected to temperatures of 130 and 180 ° for a duration of 1000 hours. The objective of the experiment was to assess the acidity level of the oil samples, measured in terms of total acid number (TAN), following the guidelines outlined in the JIS K2501 standard. The TAN value of synthetic ester oil exhibits a notable increase as temperature increases, surpassing that of silicone oil due to the superior oxidation stability properties possessed by silicone oil compared to other oils. This observation was made during the measurement of natural ester oil. This suggests that the thermal deterioration of both natural ester and synthetic ester is more pronounced throughout extended periods of thermal aging [22].

In their study, Abdul Rajab et al. conducted measurements on the dissipation factor of a palm oil sample within the temperature range of 20–100°. The measurements were carried out using the Schering bridge circuit and a null indicator oscilloscope, following the guidelines outlined in the International Electrotechnical Commission (IEC) standard 60247. The obtained results are compared to those obtained using silicone oil and mineral oil, which are frequently employed. The dissipation factor of palm oil exhibits a comparable magnitude to that of silicone oil within the temperature range of 20–60°. In the context of increasing temperatures, it is observed that the DDF (differential distribution factor) of palm oil is comparatively lower than that of SO (sunflower oil) and MO (mustard oil). This discrepancy can be attributed to the fact that the DDF is significantly influenced by the conductivity of the oil. The electrical conductivity of an oil sample is contingent upon the presence of dissociated molecules within the sample. On the other hand, it should be noted that palm oil exhibits a high viscosity, leading to a limited movement of ionized molecules. Consequently, the oil sample's diffusion coefficient is relatively low, as indicated by previous research. In their study, Abdul Rajab et al. conducted measurements on the dielectric constant of a sample of palm oil throughout the temperature range of 20–100°. The measurements were performed using the Schering bridge circuit and a null indicator oscilloscope, following the guidelines outlined in the International Electrotechnical Commission (IEC) 60247 standard. The obtained results were then compared to those of silicone oil and mineral oil, which are commonly employed in similar applications. The dynamic viscosity coefficient (DC) of palm oil is recorded as 3.26 at a temperature of 25 ° and 3.23 at a temperature of 100°. In comparison, the DC values for mineral oil and silicone oil range from 2.21 to 2.14 and 2.56 to 2.49, respectively. Based on the findings of the experimental research, it can be concluded that the dynamic viscosity coefficient (DC) of palm oil is much greater in comparison to other oils. The ease of dipole production on palm molecules, in comparison to mineral oil and silicone oil, renders them more susceptible to polarization when subjected to an electric field [23].

Both coconut oil and mineral oil were subjected to a thermal and electric test by Matharage et al. After seven weeks of heating oil samples to 120 °C, the dissolved gases are analyzed with a Myrko faults gas analyzer. The results show that compared to mineral oil, coconut oil has a slightly greater concentration of H_2, CO, and CO_2. Twenty separate breakdown experiments are performed on coconut oil, and while it produces slightly more H_2 than mineral oil, it shows off a massive amount of CO instead. During a simulated partial discharge defect, coconut oil emits a large amount of carbon monoxide and carbon dioxide, while mineral oil emits relatively little carbon monoxide and a much larger amount of carbon dioxide. Coconut oil emits more carbon monoxide and carbon dioxide during thermal and electrical faults than mineral oil does. The higher the concentrations of CO and CO_2 are, the greater the solubility. Consequently, coconut oil was more soluble in water than mineral oil. Oil samples with pressboard and metal additives produce substantial CO and CO_2 emissions [24].

Yu Liu and Jian Li did a DGA on 90–800 °C worth of mineral oil, FR3, and camellia oil. The major gas indicator of a thermal defect forming on FR3 (C_2H_2 and H_2 in the case of camellia) is C_2H_6, and up to a temperature rise of 300 °C, both camellia and FR3 created a large amount of C_2H_6 relative to mineral oil. Above 300 °C operating temperature, considerable CH_4 followed by C_2H_4 is working as a primary gas indicator of high thermal faults. During low thermal faults, mineral oil generates a large amount of H_2 and CH_4, which act as a lower thermal fault indicator. Partial discharge on Ricinus Oil was simulated by Umarkhayam et al. for DGA. Partial discharge happened in Ricinus Oil at 18 kV, but in mineral oil it only occurred at 17 kV (using a lap configuration). TCG of Ricinus oil is 16.1, but that of mineral oil is only 15.7 at the time of partial discharge, which is still below IEEE regulation. Ricinus oil has a C_2H_2/C_2H_4 gas concentration ratio of 0.71 and a C_2H_4/C_2H_6 gas concentration ratio of 1.2. The overall combustible gas generation is larger in Ricinus Oil than in mineral oil, apart from CH_4, although its value is in acceptable limits. This can be inferred from the fact that the generation of C_2H_4, C_2H_6, and C_2H_2 in mineral oil is slightly lower than that in Ricinus Oil [25].

In order to estimate the amount of carbon oxides, Ivanka Hohlein-Atanasova did a DGA on a transformer and a tap changer, with oil samples placed in headspace visala with air ingress flow at 150 °C for 164 hours. Both with and without the addition of catalysts, the synthetic ester oil shows extremely high concentrations of carbon monoxide and carbon dioxide, while the natural ester oil shows much lower concentrations of these gases. The amount of carbon monoxide (CO) released by the unrestricted oil was somewhat higher than the other samples, but the oil treated with catalysts released significantly less carbon dioxide (CO_2). In comparison to other samples, natural ester oil has about 5000 ppm higher concentrations of C_2H_6. The CO_2/CO ratio of transformers is larger in the generation unit than in the stations [26].

Partial discharge, arching, and overheating faults on biodegradable oil and mineral oil modeled for DGA by Muhamad et al. Point–point and point–plane electrode combinations were used to generate the partial discharge fault in both vegetable oil and mineral oil. Vegetable oil produces a lot of hydrogen gas and carbon monoxide at the moment of partial discharge when using point-to-plane electrode combinations, while mineral oil just produces hydrogen gas. When using mineral oil and a plane-to-plane electrode combination, acetylene is produced at a rate 33 times higher than the standard limit. When put through an arching test, biodegradable oil produces over 200 ppm of H_2, while mineral oil produces 150 ppm of acetylene. While both types of oils produce some C_2H_6 and H_2 during an overheating fault, the amount of dissolved gases produced by biodegradable oil is significantly higher [27]. Gomes and Augusta G. Martins oil samples were processed through DGA with and without pressboard to determine if they were made from natural or synthetic ester. Synthetic ester oil produces more carbon monoxide and carbon dioxide than natural ester oil or mineral oil does during the aging process. Mineral oil shows the lowest amount of H_2 at all temperatures over 100 °C, while synthetic ester and natural ester oils show a substantial increase beyond 110 °C of operating temperature. All oil samples show a negligible quantity of CH_4 at temperatures up to 70 °C. Synthetic ester oil produces significantly more carbon monoxide (CH_4) at 130 °C than other samples. Mineral oil emits negligible amounts of C_2H_4 at all temperatures, while natural ester emits more than twice as much at temperatures over 150 °C and synthetic ester oil emits more than twice as much at temperatures below 110 °C. While other oil samples indicate relatively low amounts of C_2H_4, the emission rate of C_2H_6 in natural ester has grown dramatically with the rise in temperature. All oil samples with pressboard emit much more carbon monoxide and carbon dioxide than those without. Carbon monoxide and carbon dioxide levels are highest in synthetic ester oil combined with pressboard, followed by mineral oil and natural ester with pressboard [28].

Soy-seed-based oil and mineral oil were subjected to a DGA with and without pressboard at varying moisture levels while partial discharge, arcing, and overheating faults were simulated. Soybean oil generates less hydrogen, methane, ethylene, and acetylene than mineral oil does at the instant of PD faults across all sampling settings. Since the chemical bonds in oil derived from soybeans are much stronger than those in mineral oil, only a little amount of hydrogen and hydrocarbon gases are produced during processing. Soybean-based oil produces 2337 ppm more acetylene (at wet) than mineral oil (565 ppm), 879 ppm more hydrogen (at wet), and 161 ppm more hydrogen (at dry) than mineral oil. Soybean seed oil produces more ethane (398 ppm at wet) than mineral oil produces methane (151 ppm at normal) at the instant of overheating fault, but the reverse is true for methane generation. When pressboard is combined with a mineral oil sample, the amount of all flammable gases produced after partial discharge is double that of a soy-seed-based oil. Pressboard samples impregnated with mineral oil show twice as much acetylene during an arcing fault as those impregnated with soya-based seeded oil. When compared to a soybean impregnated pressboard oil sample, mineral oil produces much more of every gas except acetylene. Soybean seed oil impregnated pressboard samples show twice the formation of ethane as mineral oil during overheating faults [29].

Toepler pump extraction per IEC 60567 and headspace extraction per ASTM D3612-2 were used to remove dissolved gases from the distribution transformer, and the amount of gas removed was compared. Both methods produce the same amount of carbon monoxide (CO) up to 43 years of age, but after that, the headspace approach produces more CO than the Toepler pump does. With the exception of C_2H_6 and C_2H_2, the concentration of all other flammable gases is rising linearly with time. dissolved gases are produced at a higher rate by older transformers than by newer models. The insulating medium's weak chemical bonding is to blame for this problem with an older transformer. The Toepler approach results in lower concentrations of all gases compared to headspace except for CO_2 and C_2H_2. Because of this, only a fraction of the gases present in oil samples are recovered using this technique. Variations in flammable gas production are also found between transformer models [30].

Factory testing, installation, and operation were all subjected to DGA by Daniel Martin and company. Combustible gas levels do not rise during factory testing, suggesting that the transformers are made to specification. Prior to the temperature increase and overload tests, mineral oil was subjected to DGA. DGA data show that whereas H_2, C_6H_6, CH_4, C_2H_4, C_2H_2, and CO all have concentrations below 1 ppm, CO_2 is at 194 ppm, O_2 at 1.169 ppm, and N2 at 4492 ppm. After an HV test was performed on the oil sample, the concentrations of C_6H_6, CH_4, C_2H_4, C_2H_2, and CO were unchanged, but the concentration of H_2 reached 1 ppm because bubbles can form during the test, greatly increasing the concentrations of CO_2, N_2, and O_2 [31].

Dissolved gas in oil samples from the Gemini X and FR3 was analyzed by Wang et al. Gas chromatography is used to determine the amount of gas in both the online approach using the TM8 DGA monitor and the offline method

TJH2b according to IEC 60567. Heating elements were used to simulate a thermal fault on oil samples, with the element being heated to 400 °C and held at that temperature for 5 minutes. C_2H_4 and CH_4 are produced in large quantities at the moment of thermal fault and serve as important gas fault indicators. Hydrocarbon gases account for 11.5% of the total measured volume, although oxygen and carbon monoxide make up 60% of the volume. Since there is a lag time of 8 hours between the online and offline approaches, there is time for absorption and leaking. Until breakdown occurs, oil samples will have their applied voltage increased by 1 kV/s. When compared to thermal faults, the number of gas releases was negligible. More hydrogen and carbon dioxide than hydrocarbon gases were created. The increased wait time between online and offline techniques accounts for 20% of the disparity. There is no continuous arching generated within the oil sample at the time of fault sparking; hence, the production of hydrocarbon gases is minimal, with the exception of C_2H_2. Since the offline measurement process takes 16 hours longer, the difference between the two methods is 19%. Based on the findings above, it is clear that syringes are necessary for timely transmission between transformers in lab scale tests and for effective scaling. A literature review on early problem diagnosis in power transformers utilizing DGA was conducted by Sukhbir Sighn et al. from 1975 to 2008. Hydrogen and methane are produced at temperatures of about 150 °C, ethane at 250 °C, ethylene at 350 °C, and acetylene at temperatures between 500 and 700 °C, according to the survey. Cellulosic materials degraded over 100 °C, releasing CO, CO_2, H_2, CH_4, and O_2 [32],[33].

Natural esters FR3, Midel eN, synthetic ester model 7131, and mineral oil Lyra X were all tested for their gassing characteristics under thermal and electrical stress according to IEC and ASTM standards by Jovaleki et al. A lap arrangement produces the 134-kV breakdown voltages that are used to test oil samples. A gas chromatograph is used to collect and analyze all gases, even those that are not free. After simulating 90 breakdowns on oil samples, researchers found that mineral oil produced the highest concentrations of dissolved gases (C_2H_2: 100 L/L; H_2: 17,75 L/L) compared to the other oils. Various voltages (40 and 50 kV for FR3, 53 kV for mineral oil, and 35 kV for synthetic ester) were used to induce a partial discharge. Hydrogen was created at the time of the partial discharge in disproportionately large quantities relative to other gases. There appears to be more PD action on natural ester oils due to the presence of dissolved gases. Under different sampling conditions, heating FR3 oil samples (at different temperatures and time intervals) greatly increases the formation of combustible gases on FR3 oil, with the exception of C_2H_2, which is less than 1 L. Mineral oil, on the other hand, produces a lot more flammable gas as the temperature rises. The amount of flammable gas produced by heating FR3 fluid to 120 °C for 16 hours is 2028 L [34],[35].

The mineral oil Nynas Nytro 10GBN, the synthetic ester 7131, and the natural ester FR3 were subjected to simulated thermal and electrical faults by Imad-U-Khan et al. The oil samples were first dried, and then they were heated at 90, 150, and 200 °C for 14 days, to remove any remaining moisture. Mineral oil and synthetic ester oils produce more than 90% of CO at 90 °C operating temperature throughout a 14-day sampling period, while natural ester oil produces less than 20% of CO. Natural esters produce over 50% hydrogen, while synthetic ester oils and mineral oils produce less than 10% hydrogen. Furthermore, both mineral oil and synthetic ester oil only contain 1 ppm of C_2H_2, while natural esters have around 20 ppm. All oil samples produce above 90% CO at 150 °C. In comparison, the production of other gases is extremely low. At the time of the low energy discharge fault, all of the oil samples produced more than 40% CO and more than 30% H_2. When compared to other oils, mineral oil produces almost no carbon monoxide at fault conditions. The emission rates of CH_4, C_2H_6, and C_2H_2 are extremely low during cold corona fault, but H_2 is significantly emitted by mineral oil, synthetic ester oil, and natural ester oil. Both synthetic and natural ester oils are almost 20% more carbon intensive. Based on experimental study, more acetylene gas is produced during low arc discharge faults, while more hydrogen gas is produced during cold corona faults [36].

2.2 Impact of Nanoparticles

Oil derived from vegetables, as proposed by Zaid B. Siddique et al., serves as the foundation for the nanomaterial suspension. Choosing a mix-blend ensures that the insulating fluid's dielectric and physio-thermal properties remain optimal. Vegetable oils have a higher viscosity than mineral oils. TiO_2, a semi-conductive nanomaterial, and ZnO, a conductive nanomaterial, are chosen because of their electrical conductivity. This investigation seeks to compare and contrast the modified properties of the insulating nanofluid by evaluating the effects of these nanoparticles on a blend of natural ester oils. The experimental results show that the Gouy–Chapman–Stern layer, which forms at the interface between the oil and the nanoparticles over time, contributes to the enhanced dielectric properties of the NF samples compared to the base oil [37]. This layer is to thank for the enhanced interfacial conductivity. Conduction along the oil/nanoparticle interface in NF samples can result in a greater shallow trap density. It is possible to attribute the buildup of charge carriers at high voltages to the presence of these shallow traps. When compared to base oils, this makes the electron-trapping phenomenon in NFs even more pronounced. The improved dielectric characteristics of NFs can be attributed to the delayed de-trapping of electrons caused by these shallow traps. Dielectric and physio-thermal properties of a natural ester oil and mineral oil mixture are investigated as a result of the incorporation of TiO_2 and ZnO nanoparticles, respectively, due to their semi-conductive and conductive natures, respectively. Both forms of nanofluids, with their altered dielectric and physio-thermal properties, are compared. The results indicate that, up to a certain concentration limit, adding TiO_2 and ZnO nanoparticles improved the breakdown voltages of the synthesized nanofluids. The AC breakdown voltage was improved by 2% in pure mineral oil, 5% in pure vegetable oil, and 1% in oil blend due to the addition of semi-conductive TiO_2 nanoparticles, compared to the addition of ZnO nanoparticles. ZnO nanoparticles led to a relatively higher boost than TiO_2 nanoparticles did in the case of impulse breakdown voltage. This may be because conductive ZnO nanoparticles are able to trap electrons more quickly thanks to shallow tapping at higher voltages. While TiO_2 dispersion resulted in a minor decrease in the nanofluid's viscosity, ZnO nanoparticle addition resulted in an overall rise in the oil's viscosity, negatively impacting the nanofluid's cooling capability.

Vegetable oils are becoming increasingly popular and widely used as an effective replacement for traditional dielectric fluids in transformer applications. However, most vegetable oils are edible, restricting their widespread use. An innovative non-edible vegetable oil is proposed by Rizwan A. Farade et al. for use as an insulating fluid. The refined oil is a nanofluid based on cottonseed oil (CSO) that is resistant to oxidation. As an antioxidant, tert-butylhydroquinone was utilized. To counteract cottonseed oil's subpar dielectric and thermal properties, scientists turned to the emerging field of nanofluids. Nanofillers consisting of hexagonal boron nitride (h-BN) particles were presented as a means to increase dielectric strength and thermal conductivity at low weight fractions (0.01%–0.1%). In order to determine how long the created CSO-based nanofluids will remain stable, UV–Vis spectroscopy was used. After the nanofluids were ready, they were put through a battery of tests to determine their dielectric and thermal properties from 45 to 90 °C. Thermal properties include thermal conductivity and thermogram analysis, whereas dielectric properties include breakdown strengths under AC and lightning impulse voltages, dielectric constant, dissipation factor, and resistivity. The results of the experiments showed that the AC breakdown voltage might be increased by as much as 63.3% at a weight percentage of 0.1 wt%. The impulse breakdown voltage, however, has been increased by around 5%. The charge trapping properties of nanoparticles or EDL formed at the nanoparticle/oil interface were considered to account for these findings. CSO/nanofluids showed a slight increase in dielectric constant up to 0.02 wt% of nanoparticles, followed by a dramatic fall in dielectric constant above 0.02 wt% of nanoparticles. The varying forms of polarization in nanofluids were used to provide an explanation for these tendencies. Third, the charge trapping action of h-BN nanoparticles raised the volume resistivity of nanofluids while decreasing their dissipation factor. Measurements of thermal conductivity, thermal response analysis, and thermogram analysis all corroborate that nanofluids have much higher thermal conductivity compared to base CSO. In addition, the nanofluids' thermal conductivity increased with temperature, allowing for greater heat dissipation at higher temperatures. These improvements have been explained by analyzing the Brownian motion of nanoparticles, the interparticle interactions caused by the EDL at the nanoparticle/oil interface, and the phonon transport through the nanoparticles. The present study's findings and explanations of underlying mechanisms establish the superiority of CSO-based h-BN nanofluids as a promising option for use in power equipment and thermal energy management systems.

Liquid dielectrics serve two purposes: they insulate electrical currents and they keep the power transformer from overheating. Mineral oil (MO) is the most common dielectric material; however, it has a low breakdown voltage, flash point, and fire risk if it leaks or explodes. Therefore, studies investigate alternative, renewable natural materials that may one day replace MO-based dielectric in widespread use. In this study, we explore the dielectric properties of two natural esters: Ximenia oil (XO) and Kalahari melon seed (KMS) oil. To test their compatibility with power transformers, each is blended with recycled mineral oil at varying concentrations. Lifetime analysis is used to further verify the usefulness of these esters after their electrical and chemical properties have been examined to determine the ideal blending ratio for each. After removing excess water, both XO and KMS were found to be useful in this investigation, with the optimal ratios being 75:25 MO: XO and 70:30 MO: KMS. The two dielectrics are now usable thanks to lifetime analysis, which showed that their lifespan was expected to be longer than MO's. As a result, different ratios of natural ester oils work well for different applications. Mixed dielectrics' dielectric properties are significantly affected by their moisture level, as a larger moisture content reduces the insulator's resistance to breakdown stress. Because of this technology, the breakdown voltages of mixed dielectrics have increased dramatically. Compared to pure MO, the survival probability of MO blends with moisture-reduced XO or KMS is higher. The two mixed dielectrics are viable, as their breakdown takes much longer to occur due to an increased breakdown voltage, as proven by lifelong evaluation.

Heat transfer efficiency between single-phase oil and a nanofluid (TiO_2 nanoparticles-transformer oil) is investigated by simulating a three-phase distribution transformer in three dimensions by Bizhan Mehrvarz et al. Both models simultaneously examine the electromagnetic field in the transformer's solid parts and the heat transfer between the transformer's fluid and solid sections. It is recommended that pure transformer oil be mixed with TiO_2 nanoparticles to improve heat transfer and reduce the temperature of hot spots. TiO_2 nanoparticles in transformer oil were simulated at concentrations of 0.0005%, 0.001%, 0.01%, 0.02%, 0.04%, 0.06, and 0.1. It is demonstrated that adding TiO_2 nanoparticles to pure transformer oil improves heat transmission and accelerates the oil's natural convection. Adding 1% TiO_2 nanoparticles to the transformer oil, as demonstrated by the simulation results, decreases the temperature of the hottest point from 47.20 to 43.05 °C, which is crucial for extending the lifespan of the transformer. It is demonstrated that when the quantity of TiO_2 nanoparticles increases, the pace at which hot spots cool down slows down. A 0.001 (vol/vol) concentration of TiO_2 nanoparticles is shown to be optimal.

As transmission voltages increase, there is a greater emphasis on the transformer's insulating dependability. Oil-immersed transformers rely on mineral oil (MO) and paper for insulation. Transformer oil (TO) and oil-impregnated cellulose must have a higher insulating conduct to improve the insulation level of ultra-high voltage (UHV) transformers and reduce their size and weight. The insulation characteristics of transformer oil have been vastly improved thanks to a recent groundbreaking experiment involving the use of nanotechnology to liquid transformer insulation. In recent years, both theoretical and applied sectors have paid a great deal of interest to the study of nanofluids (i.e., NFs) based on transformer oil. To meet the challenge of compactness while maintaining high reliability, transformers require a cooling and insulating system of the highest order. Mo-based NFs are seen as potential future replacements for liquid insulation in dielectric civilization because they may provide these improved properties. When compared to MO [210, 212], NFs made with conductive NPs may reduce top-oil and hot spot temperatures by as much as 5 °C. In addition to saving space and weight, NFs have the potential to improve performance [38],[39]. Using NFs as liquid insulation could increase the gearbox voltage. Using MO-based NFs as liquid insulation as a replacement for MO may impact the longevity and operative dependability of the existing transformer system and reduce breakdowns due to insulation concerns [40],[41].

When used for an extended period of time, transformer oil loses both its thermal and dielectric qualities. The primary goal of our research is to identify a superior alternative to mineral oil. Nanoparticles (NPs) of titanium dioxide (TiO_2), zinc oxide (ZnO), and aluminum oxide (Al_2O_3) are encapsulated in pure mineral oil to create nanofluids (NFs). For a period of one thousand hours, NFs undergo accelerated multi-aging (thermal and electrical). The particular aging tank is developed to subject the oil to both electrical and thermal stressors. Breakdown voltage, water content, tan delta, flash point, viscosity, and pour point tests are used to ascertain the thermal and dielectric properties of the NFs. The specimens are tested both before and after being subjected to expedited aging. The dielectric and thermal properties of the oil were improved by the incorporation of NPs. ZnO and TiO_2 are the most impressive of the three NFs; however, conductive NPs suffer from the disadvantage of clumping. Due to the incorporation of shallow traps that caught high mobility electrons, the distortion of modified mineral oil was reduced in the case of semi-conductive NPs. Since fewer electrons were available to contribute to streamer growth, the dielectric strength of the NF increased. When compared to other NFs and mineral oil, ZnO NFs had superior dielectric and thermal properties. Breakdown strength was greatest for ZnO NPs, which were able to scatter the streamer and prevent an early breakdown compared to other NPs and virgin mineral oil. Due to their inert nature, ZnO NPs produce fewer oxides, preventing ZnO from becoming more WC and viscous and preserving its dielectric properties. Due to their superior thermal and dielectric properties, ZnO NFs can be utilized as a replacement for mineral oil, and the lowest amount of tan delta or loss factor in the case of ZnO-based NFs following accelerated aging.

Aged transformer oil with additives such as semi-conductive nanoparticles and natural and synthetic antioxidants is studied by Bakrutheen et al. To improve the oil's critical properties and get rid of the need for dangerous disposal, it is mixed with additives in various fractions of aged transformer oil. Before and after treatment, samples are tested for important parameters such as breakdown voltage, flashpoint, fire point, and viscosity using IEC and ASTM protocols. The use of semi-conductive nanoparticles and other antioxidants, according to the research, can restore the safe nature of the aging mineral oil. Regular testing for the presence of additives, especially at elevated temperatures, and replenishment of depleted levels are recommended. This method maximizes stability, cost-effectiveness, and shelf life extension. When properly maintained, used mineral oil can be used indefinitely without breaking down into sludge or becoming overly acidic as a result of oxidation. In one crucial respect, recovering the essential qualities of used mineral oil yields better results than using base mineral oil (ideal mineral oil). By strategically incorporating nanoparticles and antioxidants into transformer oil, we can improve its vital qualities and pave the path for the power transformer's performance to be maximized.

Research by Yuzhen. Lv et al. The mineral oil was modified using a wide variety of nanoparticles, each with its own unique electrical conductivity. The three types of nanofluids, together with pure oil, were subjected to IEC-mandated tests of AC and lighting impulse breakdown voltages. When compared to pure oil, all nanofluids have superior dielectric properties. Possible theories are suggested to explain why nanofluids have a higher dielectric strength. The rapid electrons in the transformer oil could be captured and converted into low mobility negatively charged nanoparticles by conductive nanoparticles with an extraordinarily low charge relaxation time constant. Nanoparticles that conduct electricity are used to "scavenge" electrons in a liquid. Shallower oil traps can be created by adding either semi-conductive nanoparticles or insulating nanoparticles. For these nanofluids, it is speculated that electron trapping and de-trapping processes in the shallow traps are the primary mechanism. The results of an experimental study looked into how three types of metal oxide nanoparticles with varying electrical conductivities affected the dielectric strength of mineral oil. The electrical conductivity of nanoparticles was not found to be directly related to their dielectric characteristics in nanofluids. (2) The insulating properties of mineral oil are enhanced because conductive nanoparticles can trap the rapid electrons and convert them into low mobility negatively charged nanoparticles, thanks to the extraordinarily low charge relaxation time constant. Nanoparticles that are semi-conductive or insulating can create oil traps with less depth. These could grab fast electrons and quickly release them due to their shallowness. In this process, the electrons' velocity slows down. Therefore, a suspension of these two types of nanoparticles can enhance the oil's dielectric properties.

Positive and negative breakdown voltages are related to NP size, as shown by the relationship established by Muhammad Rafiq et al. In order to create NFs with a 40% W/V concentration, three different sized monodisperse

Fe_3O_4 nanoparticles (10, 20, and 40 nm) were synthesized and then dispersed into insulating mineral oil. According to IEC standards, the breakdown strengths of lightning impulses in oil samples with and without NP suspension were measured. The breakdown strength of a positive impulse was seen to first climb to a maximum at a given size and then decline. Breakdown voltages of NFs of varying sizes were found to be lower than those of pure transformer oil, as measured by negative impulse breakdown. Possible mechanisms by which Fe_3O_4 nanoparticles might alter transformer oil's insulating characteristics are also examined. As a result of the investigation, the results show that the size of nanoparticles (NPs) correlates with both positive and negative impulses. Breakdown voltages for positive impulses of size [25 mm] are shown in Figure 3. Mineral oil and produced magnetic nanofluids (NFs) of four different sizes were examined for their positive and negative impulse breakdown strengths in this study. Positive impulse breakdown voltage studies showed that the breakdown strength of NFs increased up to a specific NP size and then decreased with increasing NP size. The dielectric strength of the transformer oil was greater than that of the NFs for all NP sizes under negative impulse voltages.

The loss of a distribution transformer (DT) has a significant impact on the reliability and safety of an electrical grid and is therefore a serious concern. Transformer insulation health is determined by hotspot and oil temperatures. Therefore, the insulating condition can be enhanced by managing and, if possible, lowering the temperature of the transformer oil. In this article, we employ the electro-thermal resistance model (E-TRM) to examine how substituting nano-oil for conventional oil affects the oil's temperature and the loading capacity increment (LCI) of DTs. Nanofluids with different quantities of multi-walled carbon nanotubes (MWCNTs) and diamond nanoparticles distributed in pure mineral oil (MO) were analyzed. The E-TRM method's numerical findings are first checked and validated against those obtained from actual operation of a 500-kVA DT. The impact of MWCNT, diamond, and the suggested oxidize nanoparticles on the transformer's heat transfer capacity is also studied and compared. In this work, we look into how utilizing MWCNT and diamond nanofluids affects the DT's oil temperature. To do so, we first compare and verify the results of E-TRM with the results of the high voltage lab, on the assumption that the parameters of mineral oil remain constant. The E-TRM technique is then used to determine the amount of convective heat transfer in a closed cavity for a variety of nanofluids. If the nanofluid was utilized as a coolant in the transformer, the oil temperature was determined after validating the suggested E-TRM approach at rated and varied loadings.

Then, a new equation is developed to study how the incorporation of nanofluids impacts the transformer's LCI. According to the sensitivity analysis, the thermophysical parameters, most importantly the thermal conductivity, should be determined with sufficient precision so that there is no discrepancy between the theoretical and experimental results. The results for MWCNT and diamond nanofluids show that using any nanofluid decreases the oil temperature in a transformer, and that this decrease is proportional to the load on the transformer. The volumetric concentration of MWCNT nanofluid is the lowest (0.005%), while the volumetric concentration of diamond nanofluid is the greatest (0.010%). At its rated loading, this nanofluid can produce a temperature drop of nearly 1 °C. The observed results indicate that diamond nanofluids have a higher beneficial effect than MWCNT nanofluids across the whole volumetric concentration range. Increase the concentration of nanofluid CHTC until the overall thermal resistance (OTR) becomes discernible for best results when using nanoparticles in transformer oil.

Nickel ferrite ($NiFe_2O_4$) powder samples with a range of shape and nanoparticle size have been synthesized using electrospinning, hydrothermal, green, and sol-gel synthesis techniques. The powders were characterized using XRD, FTIR, SEM, TEM, and BET methods. The ferrite that was made had a cubic spinel phase and had several morphological forms such nanofibers, nanotubes, nanorods, and nanospheres. All samples exhibited semi-conducting behavior, as measured by an increase in electrical conductivity as temperature was increased. It was discovered that the conductivity values varied with the size and shape of the nanoparticles. The results of the regress study indicate that semi-conducting behavior was shown by an increase in the DC electrical conductivity of all samples as the temperature was raised. Especially at higher temperatures, the band gap and electrical conductivity values shift due to morphology fluctuation. Potential uses for these characteristics include recording, storing memory devices, optical devices, and perhaps military use. The hopping process can be used to explain why the AC electrical conductivity of all samples improved with increasing frequency. The existence of two types of charge carriers, n-type due to the hopping of electrons between Fe_2+ and Fe_3+ ions and p-type due to the hopping of positive holes between Ni2+ and Ni3+ ions, was used to explain the conduction process. All morphological structures exhibited maximum values for their respective dielectric constants and dielectric losses at a specific frequency. The Maxwell–Wagner theory of interfacial polarization, in agreement with the Koops phenomenological theory, was successfully used to interpret the characteristic dielectric dispersion nature observed at lower frequencies. Assuming that electrical polarization is the mechanism behind the polarization process helps to provide qualitative clarity on this behavior. All of the samples have an extremely high dielectric constant. The complicated impedance analysis confirms the grain boundary's crucial role in the conduction process at extremely high frequencies. All samples tested had high AC conductivity and low dielectric loss at high frequencies, making them ideal for use in high-frequency power transformers. The effect of morphology on magnetic storage devices and medical sensors, as well as its uses in other medical sectors such antibacterial agents, immunoassays, hyperthermia therapy, and magnetic resonance imaging, will continue to be studied in future studies.

An experimental study of nano-oil's alternate insulator breakdown voltage (AC) and convective heat transfer behavior in electrical transformers is detailed in this study. Nitro libra type transformer oil, one of the most popular mineral oils, serves as the foundation for the analyzed nano-oil. Nanoscale magnetic iron oxide particles are included. After the necessary equipment has been built in the lab, convection heat transfer is studied experimentally in laminar flow circumstances with a constant heat flux applied to the wall. For alternating breakdown voltage testing, the IEC 60156-compliant BA100 breakdown voltage measurement device is employed. Considering the importance of improving the performance of oil as a cooling fluid and electrical insulation in electrical transformers, the heat transfer coefficient of nano-oil containing colloidal iron oxide nanoparticles was investigated. Thermal testing of nano-oil samples was conducted in a laminar flow inside a horizontal tube, and the results were contrasted to those of testing with base oil. To further investigate the effect of nanoparticles on the oil's dielectric properties, an alternating breakdown voltage test was performed on the samples in accordance with the IEC60156 standard. The results showed that the convection heat transfer coefficient of nano-oil with a volume concentration of 0.1% was improved by 4.51% compared to base oil. The findings of an alternating breakdown voltage test showed that the nano-insulating oil was 23.8% more powerful than the base oil. When it comes to electricity and the careful calculation of air gaps between moving parts inside the transformer, this is a vital issue.

Using a rectangular chamber, the researchers measured the breakdown voltage and thermal characteristics of transformer oil after adding either covalently functionalized-hydroxylated multi-walled carbon nanotubes (MWCNTs-OH) or non-covalently-functionalized MWCNTs. The high intrinsic thermal conductivity of multi-walled carbon nanotubes compared to other nanoparticles led to their selection for use in the present study, which is novel because it involves the use of covalently functionalized hydroxylated and non-covalently functionalized multi-walled carbon nanotubes in transformer oil. In this study, we looked at the effects of different concentrations of MWCNTs and MWCNTs-OH (0.001 wt% and 0.01 wt%) and input powers range from 50 (W) to 120 (W) on the free convective heat transfer coefficient, Nusselt number, Rayleigh number, and Gras so number under a free air flow condition. Furthermore, the role that forced air flow plays in the free heat transfer of transformer oil was studied. Covalently functionalized MWCNT-OH at a concentration of 0.001 wt% has a negligible influence on the breakdown voltage, as evidenced by the superior performance of the nanofluid with a decrease in voltage of 4.1%. With 0.01 wt% non-covalently functionalized MWCNTs, the breakdown voltage drops by 28.44%, giving us our minimal value. Covalently functionalized MWCNT-OH at concentrations of 0.01 wt% and 0.001 wt% resulted in a 26.23% and 30.08% increase, respectively, in the heat transfer coefficient for free transfer and employing a fan. Nanofluids containing 0.01 wt% covalently functionalized MWCNTs-OH additive had a greater decrease in Rayleigh number compared to nanofluids containing MWCNTs/oleic acid and base oil, indicating a high natural heat transfer. Nusselt number for nanofluids at free air flow can be maximized with an input power of 120 (W) and a concentration of 0.01 wt% covalently functionalized MWCNTs-OH, resulting in an increase of around 17.6%. Increasing the buoyancy force for the nanofluid at greater heat fluxes causes the fluid to rotate at a higher rate, which may be one technique to regulate and lessen the precipitation of nanoparticles in the nanofluid.

3

Spectroscopy Analysis

3.1 Scanning Electron Microscopy

Rui-jin Liao and colleagues used SEM to conduct a microstructural examination of thermally (unaged, two-week, and four-week aged) aged pressboard samples. The scanning electron micrograph reveals a tightly packed, intricate network of cellulose fibers in the fresh sample. Cellulose fibers are typically 40 m in width; when put in order after being thermally aged for two weeks, the average width of the fiber does not vary, but the fiber walls do undergo minor deformation. However, the primary and secondary fiber walls became disorganized after four weeks of heat aging on pressboard. However, the pressboard sample developed tiny cracks due to the exposure of fibers in the middle layer. Samples of pressboard narrowed to an average width of 20 m [42].

In order to examine the aging status of a pressboard sample using SEM, Shi-Qiang Wang and colleagues conducted an accelerated thermal aging test at 140 °C on oil-immersed pressboard samples at various aging intervals (0, 120, 250, and 400 hours). The researchers found that the new cellulose fibers have a uniform width, with an average fiber width of 30 m, and a clean, well-organized surface, in addition to superior tensile and mechanical strength. However, surface cracking and increased production of ultrafine fibers were seen in the 120-hour-old sample. The surface also got rougher and the tensile and mechanical strengths diminished as a result of this process. The average width of the 400-hour-aged sample decreased to 15 m, suggesting that pressboard samples deteriorate over time [43].

The status of solid insulating materials was tracked by Sanjeeb Mohanty and Saradindu Ghosh using SEM images while they were subjected to voltages of 0.7, 1, and 1.4 kV. Two different solid insulating materials, leatherite paper (0.13-mm thick) and Manila paper (0.06-mm thick), have been used in the study. According to the scanning electron microscopy results, the fibers in the newly collected sample are firmly packed, and the cellulose fibers are neatly arranged. Leatherite paper turns white when subjected to an applied voltage in the region of 75% of the breakdown voltage, indicating the surface is becoming rough and deteriorated. As a result, the sample is pierced when the voltage is increased to 1.4 kV. A 180 m × 100 m area has been pierced. Stress was applied to 0.06-mm-thick Manila paper at voltages of 0.4, 0.6, and 0.8 kV. At 0.6 kV, the middle of the samples turns white, indicating that the voltage is too high. The region of chain scission at 0.8 kV is 3000 m, and the puncture will emerge at the center of the sample (owing to bond breakage). According to the data, PD activity is rather crucial [44].

Using SEM, Jiaming Yan and colleagues studied the topology of three-layer insulators. The epoxy resin board device securely fastens the sample to the ground electrode composed of stainless steel. Mineral oil completely surrounds the apparatus. Then, a poly methyl methacrylate airtight container is used to keep the contents fresh. The sample was hit with a 10-kV electrical pulse from an electrode. The sample's surface started out smoothening. Some glitters and surface droplets showed up unevenly in the middle of the PD process. The surface of the fiber then develops crystalline solids, in the form of droplets. If the process of damage continues, fibrillation and a rise in crystalline solid size will follow [146].

R. Thermally aged samples of Kraft paper and pressboard were analyzed for their microstructure by Karthik et al. using SEM. The mineral-oil-soaked Kraft paper is subjected to thermal aging at 90 °C for 20 hours, while the pressboard samples are subjected to thermal aging for 100 hours. The SEM analysis reveals that the cellulose fibers in the fresh and aged samples are both tightly packed into chains without any bond breaking having taken place, respectively. The cellulose fibers' structure, specifically their side walls, is severely compromised. Additionally, the pressboard average width is lowered from 19.05 to 15.65 mm, and the average width of cellulosic fibers in Kraft paper is reduced from 19.59 to 19.35 mm [45].

R. Using SEM analysis, Karthik et al. assessed the heat deterioration of a sample of Kraft paper, pressboard soaked in ester oil. The mineral-oil-soaked Kraft paper is subjected to thermal aging at 90 °C for 20 hours, while the pressboard samples are subjected to thermal aging for 100 hours. The degradation experiments are compared to those of ester-oil-immersed Kraft paper and pressboard subjected to the identical aging conditions. SEM research shows that the margins of the cellulose fibers in 20-hour-old Kraft paper have degraded. Additionally, the average width of the fiber needs to be shrunk from 19.59 to 19.05 mm. The heat degradation of ester-oil-immersed solid insulation is less as compared to mineral-oil-immersed solid samples [46], but in 100-hour-aged pressboard samples, cellulose fibers are somewhat deformed and the width of cellulose fiber is reduced to 19.02 mm.

3.2 X-Ray Diffraction Study

Researchers Rui-jin Liao et al. analyzed the heat degradation of an old pressboard insulating sample using XRD. It demonstrates that a worn transformer pressboard emits both jagged and rounded diffraction peaks. It is assumed that transformer pressboards have both crystalline and amorphous zones. Additionally, crystalline domains are tightly packed and arranged. Amorphous areas, on the other hand, are chaotic, irregular, and prone to rapid deterioration. Diffraction peaks move as a result of crystal lattice reset on the examined specimen, which occurs naturally with age. The extended thermal aging procedure does not alter the crystal type of the pressboard sample, as evidenced by the constancy of the peak in the spectrum. The insulation of the cellulose transformer was severely degraded by age after only two weeks, with the relative crystalline decreasing from 85.98% to 75.92%. Additionally, both the crystalline and amorphous regions exhibit non-linear changes as they age. As thermal aging progressed, crystals shrank, suggesting that the pressboard in that area had been warped [42].

R. Kraft paper and pressboard are subjected to a 20-hour and 100-hour thermal aging test by Karthik et al. After that, XRD testing is done on the solid insulators. Using the contrast between the materials' ordered and disordered peaks, we can evaluate their relative intensity. In addition, the Scherer formula is used to determine the average domain size of a crystal. Crystalline domain size is determined to be 3.084 angstroms from an analysis of an aged Kraft paper sample with a relative intensity of 35.04%, while the relative intensity of mineral-oil-impregnated Kraft paper after 20 hours of aging is 14.65% and the mean crystal domain size is 1.58. The relative intensity of an aged pressboard sample is 23.44%, and the crystalline domain size is 2.16. Therefore, the relative intensity of 100-hour-old pressboard insulation is 10.54%, and the size of the crystal domain is 1.99 angstroms. Pressboard, it seems, loses some of its ferocity as it ages. According to Karthik et al., test the effect of ester oil on the durability of Kraft paper pressboards by subjecting them to 90 °C for 20 hours and 100 °C for 100 hours thermal aging test. At a location of 22.66 °C, the relative intensity of a new Kraft paper is 100%, but it remains constant in a thermally aged sample. Thus, the relative intensity of a sample after aging is 23.44%, but that of samples thermally aged at 90 °C for 100 hours is only 20.86%. There is no obvious change in the crystalline structure of pressboard or Kraft paper that has been treated with ester oil [43],[47].

3.3 FTIR Spectroscopy

The oxidation behavior of five different types of food-grade vegetable oils was assessed by IL. Hosier et al. using FTIR spectroscopy. These oils included yellow olive oil, green olive oil, rapeseed oil, corn oil, sunflower oil, Envirotemp FR3, and dodecylbenzene oil. The oxidation of different oils is compared using the absorbance at 3475 cm^{-1}. Sunflower oil has higher oxidation behavior than other oils across the board as they age. Copper's addition to oil samples hastens the oxidation of those oils. Sunflower oil and corn oil show the most oxidation among the oils tested, whereas Environtemp and green, yellow and olive oils show the least. Therefore, rapeseed oil has a moderate oxidation stability. In contrast to vegetable oils, DDB seldom ever oxidizes. Olive oil and maize oil both include carotene and chlorophyll, but these compounds are not affecting the oxidation behavior of the oils. Since olive oil contains a comparatively small amount of poly-unsaturated fatty acids, making it more resistant to oxidation, oil composition is playing a crucial role in the oxidation behavior of oil samples. In high-voltage uses, it excels [48].

A group of researchers led by Shivani Hembrom et al. FTIR spectroscopy was used on a number of field-collected and lab-processed samples to investigate the impact of age-related phenomena and contamination on insulating oil. The spectral response of the newly sampled oil reveals many peaks. The 610–725, 2840–3230, and 1420–1468 cm^{-1} equivalent wave numbers for CO bonding of CO^2 and C–H stretching of CH_4 and C^2H_4 were reported. One of the samples was taken from a transformer that had been serviced for 35 years; it tested positive for carbon monoxide, methane, and ethane. Color change in transformer oil, however, is indicative of the breakdown of paraffin, naphthenic components, and aromatic compounds, and so has a major impact on the oil's refractive mode. In addition, the absorption of transformer oil thermally aged at 60 °C for 3 hours with and without paper is nearly same. In contrast, the sample of oil to which copper was added hardly altered in its absorptions. The spectral response of the samples has not changed much even after being heated to 120 °C for 3 hours. The spectra of the samples overlap with one another. Even after 3 hours of thermal aging at 150 °C, the transmittance of an oil sample increases and its absorption level decreases. The sample of transformer oil soaked in Kraft paper has the maximum transmittance (30.016%), followed by the sample soaked in transformer oil containing copper (16.909%) and the sample soaked in fresh transformer oil (5.408%). Absorption is enhanced in Kraft paper because thermal degradation raises the oil's variance components [49].

Atikah Binti Johari et al. used FTIR spectroscopy to analyze oil samples taken during a long-term thermal aging test conducted on PFAE and Environtemp FR3. Samples of oil are heated to 150 °C for 200 hours, with FTIR measurements taken every 50 hours. Palmitic acid is released for PFAE and oleic acid is released for Environtemp FR3 as the primary fatty acids in the oil samples. The thermal aging process has a major impact on these fatty acids. Major peak shifts in oil samples were detected via FTIR spectroscopy. Over time, however, the C–H and C–OH bonds in PFAE and the C–OH and C=O functional group begin to appear in the spectral region. Due to a chemical reaction that reduces the electrical and chemical properties of insulating oil, the intensity of the peak effectively varies with the aging rate. When compared to Environtemp FR3, PFAE's resistance is on par [70].

Using FTIR spectroscopy, Fonfana et al. successfully detected age-related compounds in transformer oil. In the first stage, samples of pressboard are used to age transformer oil at 110 °C for 500, 1000, 1500, 2000, and 2500 hours. Absorption peaks are shown to grow with age via Fourier transform infrared spectroscopy. In addition, the levels of acids and peroxides rose as people got older. Oil oxidizes and thermally decomposes during aging, producing C=O and C–H. The spectral region between 1600 and 1820 cm^{-1} contains aging byproducts as aldehyde and carbolic acids. In addition, between 1450 and 1600 cm^{-1}, an aromatic compound's fingerprint is shown. In addition, peaks at 1610, 1710, and 1775 cm^{-1} are identified as those of aldehydes, ketones, carboxylic acids, and derivatives, respectively. There are carbolic acids, appearing at 1610 cm^{-1}, peaking at 1450 cm^{-1}, and aromatic chemicals in transformer oil that has been aged for 1000 hours. Addition of the pressboard to the oil sample dramatically reduced absorption, indicating that the solid components absorbed some of the acid. The pressboard in the transformer is deteriorating faster due to the acids, which in turn reduces the effectiveness of the oil cooling system. The results were most consistent when oil samples were collected on a regular basis and analyzed using FTIR [71].

E. Using FTIR Nicolet 4700 spectrometers, Rodriguez-Celis et al. pyrolyzed Kraft paper to examine the structural and chemical changes of cellulose. The O–H stretch hydrogen bonded band intensity is significantly lower than in modern Kraft paper, and two novel bands, C=C at 1590 cm^{-1} and C=O at 1700 cm^{-1}, are exhibited. The 890 and 1110 cm^{-1} glucose ring stretch bands are also still detectable. In anaerobic conditions, FTIR research shows that cellulose chars between 190 and 390 °C [52].

Using FTIR, Suleiman et al. looked into the functional group shifts in palm oil samples to gain insight into the underlying physical shifts. Three different oils, including red palm oil, refined palm kernel oil, and palm fatty acid ester, are analyzed and compared to mineral oil in this study. The effect of moisture on oil properties was also studied. The O–H bonding vibration in red palm oil occurs in the IR region of 1500−1300 cm^{-1}. The two most prominent frequencies are 1463 and 1378 cm^{-1}. The gearbox intensity drops from 86.7% to 83.77% when the moisture level is raised from normal to 0.07% but rises to 84.24% and 84.5% after adding picture content to the oil sample in the 0.2% and 0.3% ranges, respectively. In general, the number of molecules present in compounds can be inferred from their intensity, which increases as the concentration of species increases. Red palm bond energies do not follow the typical BDV distribution. In contrast, the BDV pattern corresponds to the percentage intensity. As a result, the bond energy shifts from 3600 to 3100 cm^{-1} during O–H stretching, a transition that is reminiscent of the properties of BDV in oil. The peak also follows the typical BDV shape. However, the percentage of gearbox or absorption in the IR spectrum range of PFAE oil does not significantly alter with moisture content. When the humidity is raised to 0.3%, a change in the spectra's appearance indicates that the distance between atoms is shrinking. The C–OH bonding structure, with its major peak at around 1168 cm^{-1}, is appearing between 1250 and 1000 cm^{-1} as a result of the synthesis and transesterification of fatty acid alkyl esters from acid methyl esters and alkyl alcohols. The BDV pattern is reflected in the structure of the C–OH bond. Based on these results, it may be concluded that water dilutes the strength of chemical bonds in molecular structures. Synthetic palm oils, like red palm oil, are unstable when exposed to moisture, which greatly diminishes their physical qualities [53].

3.4 UV-VIS Analysis

Using ultra-violet/visible spectrometer (Perkin Elmer Lambda 35), Hosier et al. compared the stability over time of five vegetable oils of food grade quality to that of Environtemp FR3, a dodecylbenzene oil. Yellow olive oil, green olive oil, rapeseed oil, maize oil, and sunflower oil are the vegetable oils. The spectrum is examined both before and after aging, with and without the addition of a catalytic condition. Oil is aged at temperatures between 110 and 135 °C. Dodecylbenzene oil has an absorption edge at 320 nm and is colorless and transparent before it ages. The absorption edge of sunflower oil, Envirotemp FR3, and rapeseed oils is around 400 nm, despite their pale yellow appearance. Olive oil and maize oil, which are both green and yellow, contain carotene. Two peaks at 670 and 610 nm can be seen in the spectrum, indicating the presence of chlorophylls a and b, both of which are found in green olive oil. The oil's hues become more muted as it ages. Carotene's absorption edge moves to longer wavelengths, and vice versa. After 264 hours, the strong carotene absorbance effect is significantly diminished, while the intensity of chlorophyll peaks is gradually diminished as it is gradually converted to pheophytin by the loss of its core magnesium ion. Additionally, the absorption edge of green olive oil moves from 400 to 420 nm, and the chlorophyll peaks move from 647 to 603 nm. According to the results shown above, copper speeds up the aging process. Carotene and chlorophyll, found in green olive oil and maize oil, lower the dielectric characteristics of oils. Therefore, yellow olive oil, Envirotemp oil, and sunflower oil, which have undergone extensive processing, offer superior dielectric characteristics to other oils [48].

Using ultraviolet light, Shivani Hembrom investigated the effects of oil degradation and contamination in transformers. Two different types of transformer oils are compared here. The first group of samples comes from two types of transformers, one of which has been in operation for 35 years and the other of which has oil with a high content of acetylene (S1 and S2, respectively). This data is compared to that obtained using fresh transformer oil. The second batch of transformer oil is made in the lab by putting samples through 3 hours of thermal aging at 60, 120, and 150 °C. According to UV analysis, the characteristic maximum absorption of fresh transformer oil is between 350 and 565 cm^{-1}, while that of S1 is between 350 and 365 cm^{-1}, with maximum absorption at 480 cm^{-1}, and that of S2 is between 350 and 412 cm^{-1}. Both S1 and S2 have a lot of extra noise between 300 and 350 cm^{-1} because of impurities such as very high carbon content and water, in addition to the oil samples themselves. Those that were thermally aged for 3 hours at 60 or 120 °C show negligible absorption, while those thermally aged for 3 hours at 150 °C show thermal deterioration and, therefore, absorption. Maximum absorbance has shifted in wavelength [49] because its bandwidth has expanded, and this expansion depends on the contamination rate.

Thermal decade analysis was performed on oil samples by Hasmat Malik et al. using a UV–Vis spectrophotometer on new, intermediately aged, and very old oils. ASTM D-6802 is used throughout the course of this investigation. First, spectral grade heptanes are used to set the UV–Vis to zero. Next, a glass cuvette with a 10-millimeter route length is filled with test oil and put in the instrument's reference position. In the range of 360–600 nm, the instrument can scan. The absorbance starts out at zero. The samples are then transferred to the proper sample holder. A new oil sample has been acquired, and the dissolved decay products have been removed, as indicated by the shift in the absorbance curve. After that, samples of oil that are a bit older than fresh oil are substituted. There is evidence of aging byproducts in the oil samples, as shown by an increase in absorbance value accompanied by a maintenance of the same behavior of absorbance and its wavelength. In subsequent studies, the absorption value was shown to have increased due to the filling of a very old transformer. The UV scan also reveals significant oil degradation products. The above research shows that the absorption curve moves upwards or downwards depending on the level of contamination. This technique has a high rate of success [54].

The influence of dust on samples of transformer oil was studied by Norazhar Abu Bakar et al. using ultraviolet light. At first, a scan of the cuvette containing 2.1 gm of transformer oil was carried out. Oil samples are tested to determine their absorbance and frequency range for future use as standards. Copper powder ranging from 20 to 100 mg is mixed into the oil as well. Oil samples are analyzed using spectroscopy anywhere from immediately after additive addition to 5 minutes afterwards. The absorbance maximum is less than 2 at 477.8 nm, and the fresh oil forum content is 1.75 ppm. The rate at which the additive is applied causes it to rise gradually. After adding 100 gm of copper powder, the maximum absorption level in the oil samples is greater than 2.5. The examination was repeated after a 5-minute delay, revealing that the additives had settled to the bottom of the cuvette, resulting in a considerable decrease in wavelength and maximum absorbance level. When spectral response errors are analyzed, however, it becomes clear that the influence of conductive particles on the spectral response is less than 0.01% even after a 10-minute delay. Thus, the instantaneous measurement is required for precise outcomes [55].

Neffer A. Gomez et al. used ultraviolet (UV) analysis on FR3 oil samples to determine how well the oil had aged. Distribution transformers that are new, 4 months old, 8 months old, and 7 years old all contribute to the sample for this study. In order to dissolve the oil samples, 5 mg of *n*-hexane is used. Because of the presence of saturated hydrocarbon, its absorbance band is below 200 nm, whereas the examined samples are formed at wavelengths greater than 200 nm. Absorption bands at 236.0 nm and absorbance of 2.140 have been measured in samples of fresh oil; at 4 months of aging, this number drops to 0.729; at 8 months, it drops to 0.619; and at 7 years of aging, it drops to 0.074, as measured by UV spectral analysis. Because of their C=C and C=H capabilities, these absorption bands are limited to - and n- electronic transactions. UV spectrum study concurs with the known chemical make-up of FR3 oil, confirming its structure. The chemical structure of the oil appears to be unchanged, and it follows that the chromophore properties of the oil have also been preserved. The three oil samples also went through an expedited test. The samples include unrefined oil, oil that has been heated to 110 °C for 360 hours, oil that has been heated to 90 °C for 360 hours with 5% water added, and oil that has been heated to 90 °C with 5% water added and Kraft paper. In contrast to earlier analyses, which showed four absorption bands, this one only shows two for these samples. The chromophore system of the oil samples has likely evolved, resulting in different chemical structures [56].

The condition of transformer oil was evaluated by Radha Karthick et al. using ultraviolet light. Both operational transformers and those that have been processed in a laboratory supply the oil samples. A 2-ml drop of testing oil sample is diluted to a 10-ml solution of iso-octane because the absorbance of transformer oil is higher than the spectrometer's saturation range. It is the standard solvent being used here. At first, a spectral analysis of pure transformer oil is performed, which reveals a wavelength of 265.60 nm and an absorbance of 0.397. However, the peak (0.365 nm) and wavelength (254.80 nm) of the spectral response from a sample of oil that has been serviced for 20 years are significantly smaller. A comparison of the wavelengths (266.60 nm) and absorption peaks (with respect to age and Kraft paper) reveals that transformer oil with added sulfur and sulfide has different characteristics. Aged Kraft paper has a higher absorption peak of 0.803%. However, the absorbance peak in 100 °C heated transformer oil (2.864) is greater than in the fresh sample (2.864), and the wavelength (218.00 nm) is shorter in the heated oil than in the unheated oil. Additionally, the wavelength of 268.40 nm and the absorption peak of 0.5 are both found in a 50-fold arcing oil sample. Higher wavelength (280.60 nm) and greater absorbance peak (3.493 nm) are seen in the oil samples obtained from line to line faulted. The peak absorption wavelength of 268.00 nm and the peak absorption rate of 3.643 are also similar for the overloaded transformer oil sample. Although the absorbance of a lightning-affected transformer shows a relatively unusual wavelength (320.20 nm), its peak is very high and approaches the saturation point of the measuring device. It demonstrated that the spectral response varies in terms of peak and wavelength depending on the state of the oil [57].

4

Experimental Analysis

4.1 Method of Evaluations

Sample oil rupturing strength is evaluated using the International Standard (IS) 6792-2008. Nutronics oil test kit was used for the analysis; it has two hemispherical electrodes with a 2.5-mm separation gap. The test cell was sterilized before the oil was poured inside. After that, the fresh oil sample was placed into an oil test cell to see if it contained any air bubbles. After the air was purged from the oil samples, voltage was given to them via the electrode. The voltage is increasing at a rate of 2 kV/s. Turning on the automatic test switch causes this voltage to be delivered to the oil sample automatically. A reading will be taken at the time the oil breakdown happened. Five separate measurements were taken with a 2-minute interval in between [57], [58].

Schering circuit test and null indicator oscilloscope are used to determine the dielectric dissipation factor in accordance with IS: 6103-2006. A capacitance test is performed, with the oil sample serving as the dielectric, by pouring it into three terminal test cells. The working temperature is increased to 100 °C and subsequently decreased to 90 °C; measurements are taken at both temperatures. Additionally, oil samples can be tested for their unique resistivity using this method [59], [60].

ISO 6103-2006 specifies the method for measuring specific resistance in oil samples. The standard unit is the ohms at a certain temperature, and it is defined as the resistance between two sides of a 1-cm^3 block of oil [99].

Total acid number, as defined by IS: 1448(P: 2)-2007, was used to determine the oil sample's acidity. The amount of KOH needed to neutralize the acids in 1 g of oil is expressed here [61]–[64].

Coulometrically produced Karl Fischer reagent is used to determine the moisture content of oil samples in accordance with IS: 6262-20001. It is applicable to the evaluation of oil samples with a lower water content. The Karl Fischer titration test cell is a coulometrically manufactured device into which this is introduced. In the titration cell, an electrochemical process produces the necessary titration [65], [66].

In order to determine the oil's viscosity, a red wood viscometer is used, as specified by International Standard 1448(P: 25)-1976. It is the amount of time needed for a specific amount of oil to flow under specified conditions. The orifice in the silver-plated oil cup of the Redwood Viscometer is a standard diameter. The time needed to collect the samples is recorded [67] after the oil sample was poured into the silver-plated oil cup and the hole was opened.

The Pensky–Martens closed cup test method is used to determine the flash point of oil samples in accordance with IS: 1448(P: 2)-1992. A small amount of oil was placed in a sealed brass test cup. Energy regulator increased the temperature. The test cup would be filled with flame and refilled at regular intervals. The vapor generated inside the test cup combined with air at the right moment, sparking a brief fire on the oil's surface for around a second [68].

According to IS: 6104-2006, the ring method is used to calculate the interfacial number. Force required to lift a platinum wire planner ring from a liquid with a higher surface tension than water is measured. The experiment was performed under extremely controlled non-equilibrium settings, with the goal of completing the measurement in under a minute from the onset of interference [69].

In accordance with International Standard 1448(P: 16)-2002, the density of an oil sample is determined using the hydrometer method. The DGA was carried out using a DGA gas analyzer with a multiple gas extractor in accordance with IS: 10593:2006, 9434-1992 [70], [71].

4.2 Experimental Results

The breakdown strength of insulating oil is its ability to sustain voltage stress without failing; this strength is determined by the insulating oil's physical chemical qualities, any contaminants in the oil, and the electrode configuration. Due to its high viscosity and density, which limit particle movement within the oil, VGO has a higher average breakdown strength than mineral oil. This is because VGO contains a greater proportion of unsaturated fatty acids. The dielectric dissipation factor is the amount of energy lost via an insulating fluid when an alternating electric field is applied to it. The quality of oil can be gauged using the dielectric dissipation factor as an index word. Since the bond between carbon and hydrogen is easily broken under thermal stress, mineral oil has a much higher dielectric dissipation factor than vegetable oil, as shown in Table 4.1, whereas vegetable oil has triglyceride, which is composed of fatty acids. As a result, its higher viscosity and density slow the pace at which oil molecules break apart. However, because the

Table 4.1 Comparison of critical properties of oil samples.

sl.no	Test Parameter	Standard	Limit	VGO	MO
1	BDS RH:49%	IS 6792-2008	Min. 60	67	63
2	DDF at 25 °C at 90 °C RH:50%	IS 6262-20001	0.002	0.00139 0.00156	0.005 0.009
3	MC in ppm at 25 °C at 40 °C at 90°C RH:48%	IS 13567-1992	50	17.8 13 8	10 7 5
4	SPR in ohm-cm, at 90 °C RH:47%	IS 6103-2006	35×10^{12}	12×10^{13}	10^{13}
5	Acid in mg KOH/g	IS 1448(P:2)-2007	0.1 - 5	0.5	0.01
6	Flash Point in °C	IS 1448(P:2)-1992	Min. 140	245	165
7	IFN in mN/m	IS 6104-2006	0.04	11	42
8	Density in g/cm^3 at 29.5 °C at 90 °C	IS 1448(P:16)-2002	0.8-1.6	0.924 0.872	0.85 0.76
9	KVin cSt at 27 °C at 40 °C at 90 °C	IS 1448(P:25)-1976	2 – 50	40 27 11	12 8 3

conductivity of oil is directly proportional to its dissipation factor, increasing the sampling temperature increases the aging byproducts on oil samples, which may increase the dielectric dissipation factor of oil samples, resulting in increased conductivity [72], [73].

Dielectric constant is defined as the ratio, at a particular point in time and under certain conditions, of the current density in amperes per square centimeter (A/cm^2) to the direct potential gradient in volts per centimeter (V/cm) parallel to the current flow within the sample [65]. Table 4.1 demonstrates that vegetable oil has a substantially higher specific resistivity than mineral oil. The breakdown strength and the dielectric dissipation factor are two more electrical properties that are directly related to it. High dielectric loss oil has low specific resistivity, while high breakdown strength oil has high resistivity.

It is a way to quantify how acidic insulating liquids are. The oil's acidity level is the most accurate condition gauge [65]. Since VGO contains more saturated and unsaturated (mono and poly) fatty acids, it is more acidic than mineral oil. Oil's oxidation stability is reduced because of its large concentration of unsaturated double bonds in fatty acids. In contrast, MO is made up of refined hydrocarbon molecules, which results in a slower acid generation rate than in VGO. It is worth noting that mineral has the LMWA while VGO has the HMWA. Although it does not react with paper, LMWAs are known to hasten the aging process [19, 18].

Water degrades cellulosic materials, copper, and transformer metal parts [27], [28], and has a significant impact on the electrical properties of insulating oil. Table 4.1 demonstrates that vegetable oil has a larger moisture content than mineral oil across the whole temperature range of the samples taken; however, this does not have any effect on the electrical properties of VGO. Vegetable oil's relative water content is much lower than that of mineral oil, which affects the oil's electrical properties [18].

It is a quantitative representation of the fluid's permeability. It is the fluid's density divided by its viscosity [19]. Oil's ability to cool is highly sensitive to its viscosity. Oil with a high viscosity hinders both its fluidity and its capacity to transport heat. Oil's viscosity increases as its age-related byproducts and oxidations accumulate [18]. Table 4.1 demonstrates that at all temperatures tested, the viscosity of VGO is greater than that of mineral oil. The viscosity of vegetable oil, like that of mineral oil, dropped as the oil temperature rose.

The fire point at which the vapor pressure is high enough to mix with air at the liquid's surface and cause combustion [22]. The amount of volatile contamination in the insulating oil can be estimated from its flash point. More volatile contaminants can be found in substances with a lower flash point [22]. VGO's flash point exceeds both the minimal requirement of IS 1448 (P: 2)-1992 and the value for mineral oil (see Table 4.2).

Interfacial tension: It is a quantitative representation of the oil-and-water molecular attraction force at the contact. Soluble polar contaminations, oil degradation, and oxidation products can all be detected with this method. According to Table 4.1, VGO has a greater acidity level than mineral oil. Therefore, it has a larger concentration of polar components than mineral oil. Therefore, VGO had lower interfacial tension than mineral oil at the same temperature of sampling. Oil density is the ratio of oil to water in a given volume. This determines insulating oil's chemical composition. Oil grade can also be measured by density. Table 4.1 shows that VGO density decreases from 0.924 g/cm^3 at 29.5 °C to 0.872 g/cm^3 at 90 °C and mineral oil density decreases from 0.85 to 0.76. This shows that VGO oil has somewhat higher density than mineral oil, although it decreases as sampling temperature increases [68]–[72].

Dissolved gas analysis of dissolved gases in insulation solutions is the cheapest, most accurate transformer status prediction approach. Insulating oil degrades when electro-mechanical stresses surpass its permissible limit, producing flammable gases like hydrogen (H_2), methane (CH_4), ethane (C_2H_6), ethylene (C_2H_4), and acetylene. The carbon-based hydrocarbon molecules in oil samples contain a lot of CO and CO_2. The VGO has less CO than mineral oil but more CO_2 [74].

4.3 Effect of Thermal Stress on Transformer

According to the findings of the experiments, the average BDV (breakdown voltage) of the fresh oil sample is low, as shown in Table 4.2 (55.6 kV for NEO and 46.7 kV in MO). This is due to the fact that the fresh oil sample has a larger moisture content than the samples of the other types. On the other hand, as can be shown in Table 4.2, the average BDV of a NEO sample that is roughly 180 days old is 61 kV, whereas the MO has a BDV of 59 kV. The moisture concentration of the oil sample is significantly decreased as a direct result of the thermal aging process. As can be seen in Table 4.2, the average BDV for a catalyst-added NEO sample is 69 kV, which is significantly greater than that for MO. This is due to the fact that MO is composed of hydrocarbon components that have distinct compositions. Because of its reaction with oxygen, this oil generates carbon monoxide, carbon dioxide, hydrogen, and sludge, all of which contribute to a significant reduction in breakdown strength. In contrast, NEO oil has a greater quantity of unsaturated fatty acids. As a consequence of this, the viscosity of NEO is higher in comparison to MO; this viscosity restricts the mobility of particles inside the oil [70]–[72].

As can be seen in Table 4.2, the dielectric dissipation factor of both of the oil samples has grown as a result of the passage of time. Because of the higher temperature, the oil molecules are able to more easily break apart, which results in the oil having a lower viscosity. Because the oil molecules

Table 4.2 Electrical properties.

Aging in days	**DC**		**DDF**		**Specific resistivity**	
	VGO	MO	VGO	MO	VGO	MO
0	4.0	2.84	12.0	14.8	20	10
10	3.9	2.84	14.0	15.8	20	10
20	3.8	2.61	16.0	19.8	20	8
30	3.8	2.58	17.0	21.8	20	8
40	3.6	2.42	18.0	23.8	19	8
50	3.6	2.3	21.0	26.8	19	9
60	3.3	2.18	23.0	29.8	19	8
70	3.2	1.98	25.0	33.8	16	6
80	3.2	1.95	31.0	36.8	16	6
90	3.2	1.95	35.0	39.8	16	8
100	3.2	1.65	37.0	42.8	14	9
110	3.0	1.65	37.0	42.8	10	7
120	3.0	1.65	37.0	42.8	14	7
130	3.0	1.65	37.0	42.8	10	4
140	2.8	1.65	37.0	49.8	10	4
150	2.8	1.65	38.0	51.8	9	4
160	2.8	1.65	38.0	53.8	9	4
170	2.8	1.65	38.0	55.8	10	4

have been dissociated, they are able to move around more freely inside the oil, which contributes to an increase in the conductivity of the oil samples [68]. Because MO are mostly made up of hydrocarbon molecules with 15–40 carbon atoms per molecule and also because these molecules are constructed of C-C and C-H bonds, the DDF of MOWC is consistently higher than that of NEOWC, regardless of the temperature at which the samples were collected. The chemical bonds that hold the atoms together can be disrupted as a result of heat stress [101]. On the other hand, it has the structure of a triglyceride in NEO. Due to the fact that it contains more than 70% unsaturated fatty acids,

Table 4.3 Chemical properties.

Aging in days	Flash		Viscosity		Interfacial tension		Density	
	VGO	MO	VGO	MO	VGO	MO	VGO	MO
0	242	170	35	9	10	41	0.814	0.35
20	240	163	22	9	10	40	0.814	0.34
30	238	160	17	9	10	37	0.814	0.332
40	237	158	15	7	10	34	0.805	0.331
50	231	155	15	7	9	31	0.806	0.39
60	227	147	12	7	9	29	0.818	0.39
70	223	145	12	7	9	27	0.812	0.41
80	214	138	12	7	9	27	0.811	0.415
90	209	134	12	7	9	27	0.807	0.3567
100	206	120	15	11	8	23	0.771	0.3567
110	201	120	15	11	8	23	0.761	0.3567
120	198	120	15	11	8	23	0.749	0.3567
130	184	110	17	11	8	17	0.752	0.3567
140	174	110	17	11	8	17	0.752	0.363
150	174	110	17	11	8	13	0.772	0.361
160	154	100	17	12	8	10	0.752	0.361
170	154	100	17	12	8	10	0.752	0.361
180	154	100	17	12	8	10	0.752	0.361

the viscosity of NEOWC oil is significantly higher in comparison to that of MOWC as shown in Table 4.3. Because of this, the movement of dissociation of oil molecules within the oil is restricted, which results in a decrease in the conductivity of the oil. As a direct consequence of this, the DDF of the oil also decreases.

The NEO has a polar character, whereas the MO has a polar alkane molecule, which affects the dielectric constant of the oil samples. Both of these factors contribute to the dielectric constant. The dielectric constant of NEOWC is significantly greater than that of MOWC, as shown by the experimental study. With increasing age, the dielectric constants of both of

Table 4.4 Physical properties.

Aging in days	Acidity		Water content	
	VGO	MO	VGO	MO
0	1.02	0.005	45	26
20	1.62	0.006	55	36
30	2.02	0.036	75	21
40	2.62	0.037	91	33
50	3.02	0.051	119	18
60	4.02	0.081	139	20
70	4.22	0.081	164	16
80	5.02	0.121	167	16
90	5.22	0.181	143	16
100	5.82	0.221	130	14
110	6.22	0.281	116	14
120	6.62	0.421	105	14
130	6.57	0.481	85	14
140	7.22	0.501	78	14
150	9.12	0.531	55	14
160	9.78	0.531	65	14
170	10.22	0.531	65	14
180	12.22	0.531	65	14

the oil samples become less stable. Because it lessens the oil's density and viscosity, it causes an increase in the number of samples. Additionally, because it lessens the dipole orientation of the oil samples, it causes a decrease in the dielectric constant [64], [72].

As the temperature continues to rise, the specific resistivity of the oil continues to drop off progressively. There is a one-to-one correspondence between it and the interfacial tension as well as the dielectric loss factor [73], [74]. When the resistivity of the oil is low, it will have a higher conductivity and will have a lower breakdown strength. As can be seen in Figure 3.6, the resistivity of the NEO is significantly higher than that of the MO. This indicates that the contamination rate of the NEO is relatively low when compared to that of the MO for the same sampling periods.

As can be seen in Table 4.3, the acid formation rate in an oil sample that has NEOWC added to it is significantly higher than that of a MOWC oil sample during the aging process. Due to the fact that the acid formation phenomenon in NEO occurs in a different manner than in MO, it does not affect the aging process of paper. In MO, acid generation follows chain off, chain continuity, and chain breaking out rather than the site pattern seen in NEO. Therefore, MO contains LMWA such as formic, acetic, and levulinic acids, but NEO contains a greater quantity of oleic acid and linoleic acid, both of which are classified as HMWA. These acids do not hasten the rate at which the paper will age, but the LMWA will react with the paper, and this will alter the rate at which the paper will age [75], [76].

The amount of water present is the factor that has the most significant impact on the critical properties of insulating oil. As can be seen in Figure 3.8, the findings indicate that the water content of the NEOWC sample is significantly more than that of the MOWC sample at each and every one of the sampling intervals. The water content of the NEOWC sample saw a considerable increase over the first 120 days of its age; after that point, it began to drop with each additional age day. After a sampling interval of 50 days, the water content of this sample, like that of the MOWC, is at its maximum. The saturation limit of NEOWC is significantly larger than that of mineral oil because of the presence of moisture. The cellulosic insulation that was employed with the oil sample will begin consuming the water that is present in the NEOWC once it reaches the moisture saturation limit in order to keep the equilibrium state stable. As a result, after 120 days of sample intervals, the amount of water in NEOWC was determined. Even though the water content decreased, the electrical characteristics that were evaluated for NEOWC were found to be superior than those of mineral oil. This is due to the fact that mineral oil had a higher relative moisture content than NEO [78], [79].

There is a correlation between the average molecular weight and the viscosity of oil, as demonstrated by the Mark–Houwink–Sakurada equation [101]. The NEOWC value is significantly more than the MOWC value. The viscosity of the oil is being significantly impacted by the aging factor as well as severe oxidation. When compared to MOWC, NEOWC consistently has a higher viscosity during the whole aging process [80]. As can be seen in Table 4.2, the viscosity of NEOWC and MOWC samples does not remain steady over the first 90 days of the sampling process. Additionally, due to the fact that the thermal aging process may enhance contaminations such as dust and metal ions on oil samples, the viscosity of oil samples is somewhat increasing.

The results of the experiments reveal that the flash point of NEOWC is 245 °C, and that it drops to 110 °C at the conclusion of the aging period. On the other hand, the flash point of MOWC drops from 160 to 90 °C, as shown in Table 4.2. The conclusion that can be drawn from this is that the aging process results in a significant drop in the flash point of oil samples [97]. Because of the accelerated aging process, however, the flash point decrease rate in MOWC is greater than that of NEOWC. This is due to the fact that the LMWA in MOWC is increased, which in turn increases the LMWA. This LMWA causes a reaction between the cellulosic insulation material and the oil, which speeds up the thermal deterioration of the cellulose insulation material. Because the HMWA that is produced in NEO does not react with the solid insulating materials, the amount of thermal aging byproduct that is produced by NEO samples is less than that produced by MO samples [81].

It is possible to detect soluble polar contaminants, oil degradation products, and oxidation byproducts by measuring the interfacial tension. The NEO contains a greater concentration of acids than the MO. In addition to this, the percentage of polar material is higher than the MO. As a result, the interfacial tension of NEOWC is significantly lower than that of MOWC during the same sample times [82]. As can be shown in Table 4.3, the interfacial tension of NEOWC drops from 11 to 9 mN/m, but the interfacial tension of MOWC decreases from 42 to 12 mN/m. As a result, the interfacial tension drops as a result of aging, which causes thermal degradation in both the oil and the products of cellulosic insulation.

When compared to MOWC's density, NEOWC's density is initially found to be greater. The density of NEOWC was unstable over the first three and a half months of the sampling period. After a few days, it started to gradually drop with the decrease in the aging period, which eventually achieves a stable value after 140 days of sampling period, whereas the density of MOWC starts to get decreased for the first 40 days of the sampling period.

Table 4.3 demonstrates, once more, that the density of NEOWC steadily increases with the increase in aging duration up to 90 days. Inducing thermal breakdown of the insulating oil together with the paper and the metallic elements is one of its effects. The density continues to steadily decrease throughout the course of additional aging periods, until it ultimately achieves a value that is more or less stable [83].

4.4 Spectroscopic Analysis

4.4.1 The X-ray diffraction study

Crystal structure and crystalline orientation are important factors in determining the electrical characteristics of cellulosic insulating materials. X-ray diffraction (XRD) research is a powerful tool for studying the cellulose fiber crystal structure in transformer solid insulation. Dimensions like length, breadth, and height are analyzed together with more obscure ones like diffraction angle, crystal structure identification, and chemical phase angle [56], [69].

Figure 4.1 displays a typical XRD pattern for a solid insulator. There are two distinct sorts of segments visible in the design. Due to the increased concentration of cellulose in paper, the diffraction pattern exhibits strong diffraction peaks with a high relative intensity associated with crystalline nature. As a result, weak, smooth diffraction peaks can be seen in the spectral data. The XRD spectrum shows that there are crystalline and amorphous zones present in the solid insulation, which makes sense given that paper contains trace amounts of hemi cellulose and lignin. In contrast

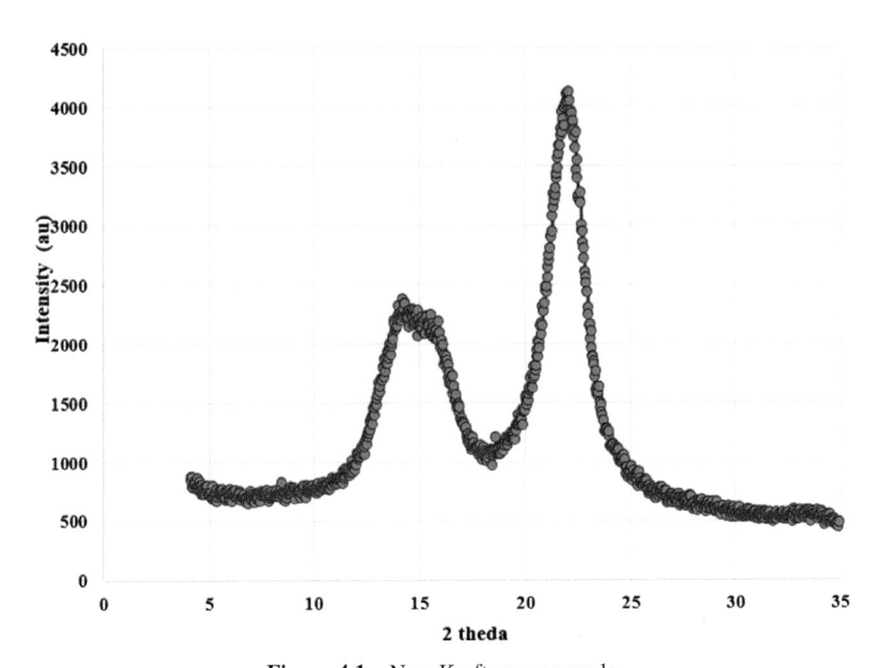

Figure 4.1 New Kraft paper sample.

to the amorphous zones, which are disorganized, irregular, and prone to deterioration, crystalline regions are compact [56], [69].

Figure 4.2 shows the XRD pattern of thermally stressed Kraft paper with mineral oil, and Figure 4.3 shows the XRD pattern of Kraft paper aged with NEO. Figure 4.4 shows the XRD pattern of new pressboard. Figure 4.5 shows

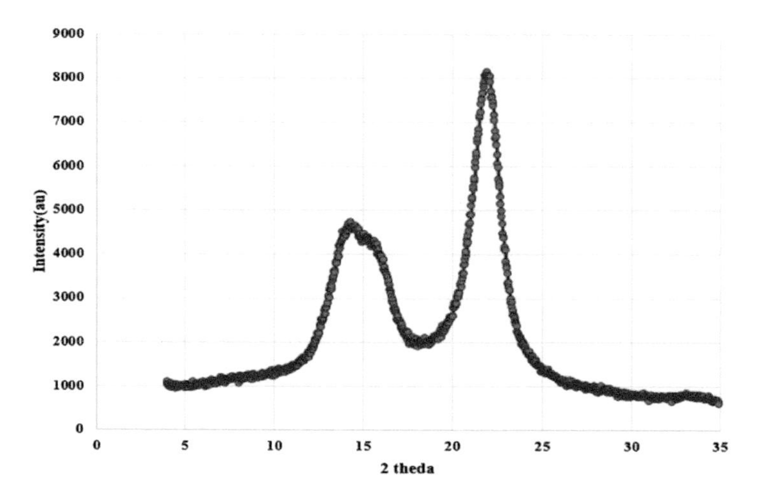

Figure 4.2 Thermally stressed Kraft paper with MO.

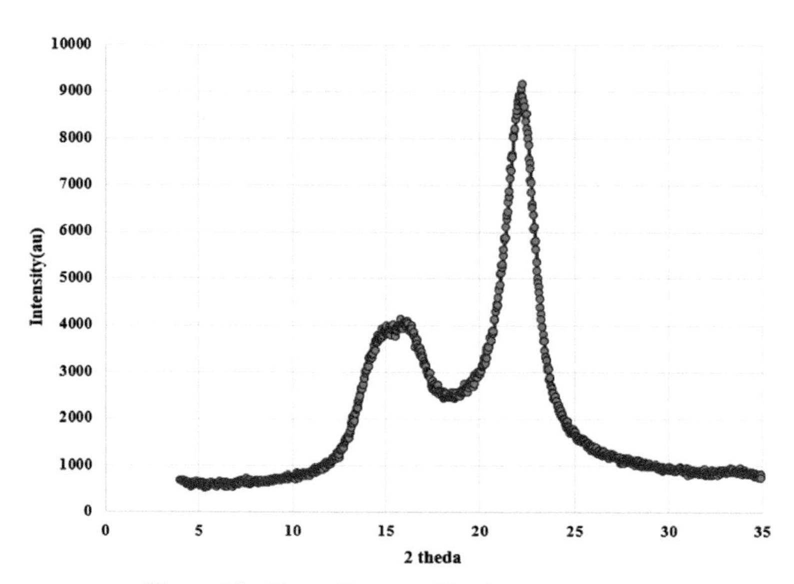

Figure 4.3 Thermally stressed Kraft paper with NEO.

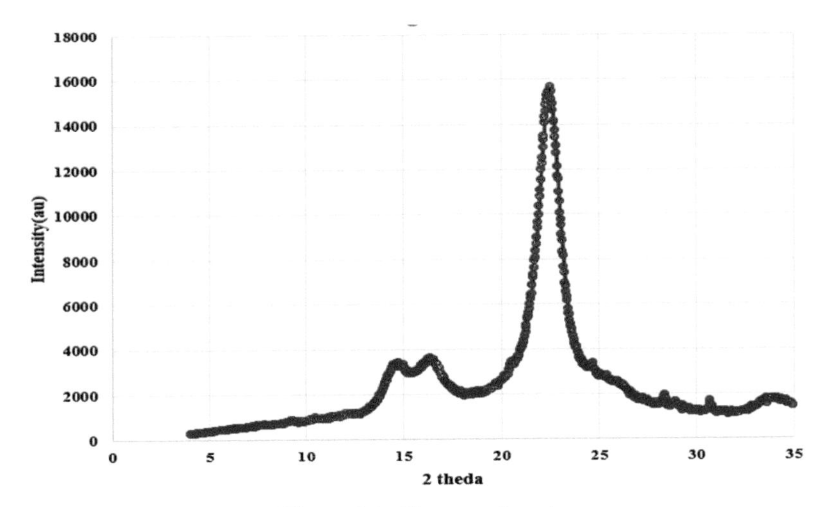

Figure 4.4 New pressboard.

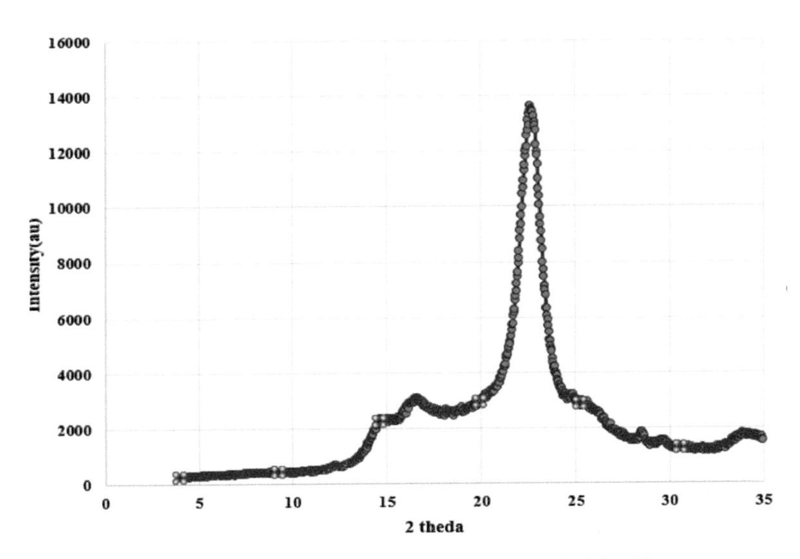

Figure 4.5 Pressboard thermally stressed with MO.

the XRD pattern of MO, and Figure 4.6 shows the thermally aged pressboard with NEO.

Insulating materials can be evaluated for alterations in terms of crystallite size, relative crystallinity, and relative intensity.

$$Dhkl = k\lambda/(\beta cos\theta), \ldots\ldots.. \tag{4.1}$$

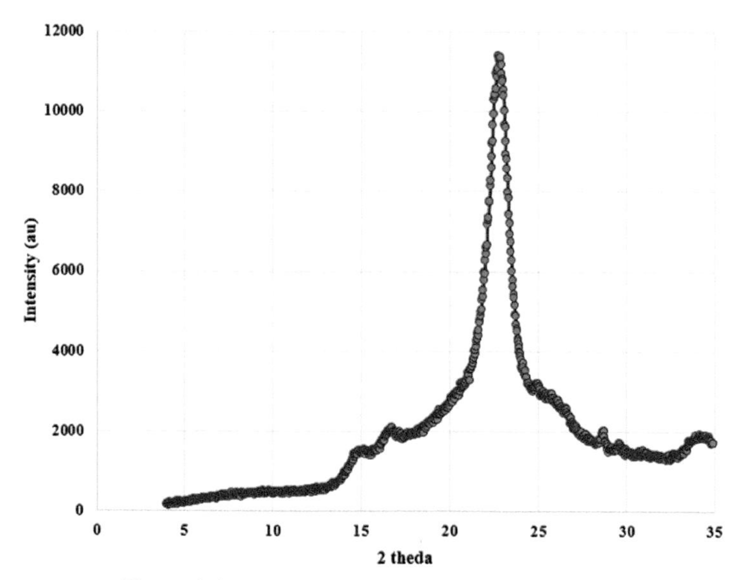

Figure 4.6 Pressboard thermally stressed with NEO.

where k is the Scherrer constant, λ is the X-ray wave length, β is full width at half maximum of the reflectance surface, and hkl is measured in 2θ (θ is the corresponding Bragg angle).

The relative crystallinity of cellulosic insulating materials is calculated using the following formula:

$$\%CrI = [(I002 - Iam)/I002] \times 100, \ldots.. \tag{4.2}$$

where CrI is the relative crystallinity, $I002$ is the diffraction intensity of crystalline region, and Iam is the diffraction intensity of amorphous region.

Because chain scission is much lower in the area of crystals versus the amorphous region, these parts have larger amounts of free energy than the crystal region. As a result, the crystallite size of the sample aged with mineral oil immersed cellulosic insulating materials is slightly reduced when compared to new cellulosic insulating materials. As a result, temperature has a greater impact in the amorphous region than the crystalline region. In contrast, when solid insulating materials are submerged in vegetable oil, the crystallite size of Kraft paper somewhat increases while the crystallite size of pressboard slightly decreases [72]–[78].

Figure 4.2 demonstrates that when solid insulating materials are submerged in mineral oil, the crystallinity index decreases by a small proportion compared to a freshly sampled equivalent. However, vegetable oil has a significantly higher proportion of crystallinity than other oils as shown in Figure 4.3. This suggests that insulating materials preserved in vegetable oil degrade at a slower rate than those preserved in other oils [76].

It is noticed that the crystallinity of thermally stressed insulating materials is decreased compared to that of the fresh sample. The crystallinity reduction rate is the best indicator for the deterioration of solid insulating materials. From the XRD analysis, it is concluded that solid insulating materials immersed in natural ester oil have higher crystallinity than others.

4.4.2 SEM

The SEM picture of unprocessed Kraft paper is displayed in Figures 4.7 and 4.3 at ×200 and ×400 magnifications, respectively. It reveals intertwined and ordered bundles of cellulose fibers. Additionally, the cellulose fibers do not separate from one another because of bond breakage. In addition, cellulose fibers organize into chains [47], [69]. A scanning electron micrograph (SEM) image at ×200 magnification of aging Kraft paper in mineral oil is shown in Figure 4.8. It demonstrated that there was no major disruption in the normal sequence of cellulose fibers. Figure 4.9 shows that this same area,

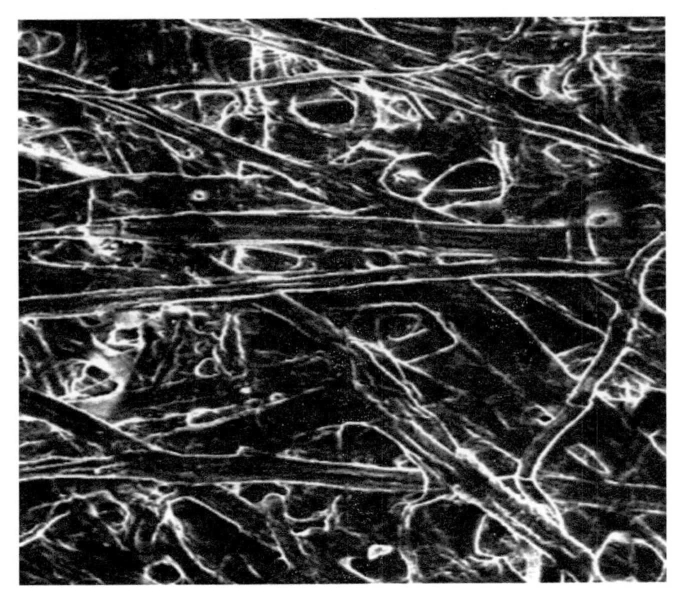

Figure 4.7 SEM of Kraft paper.

Figure 4.8 SEM of Kraft paper thermal stressed with MO.

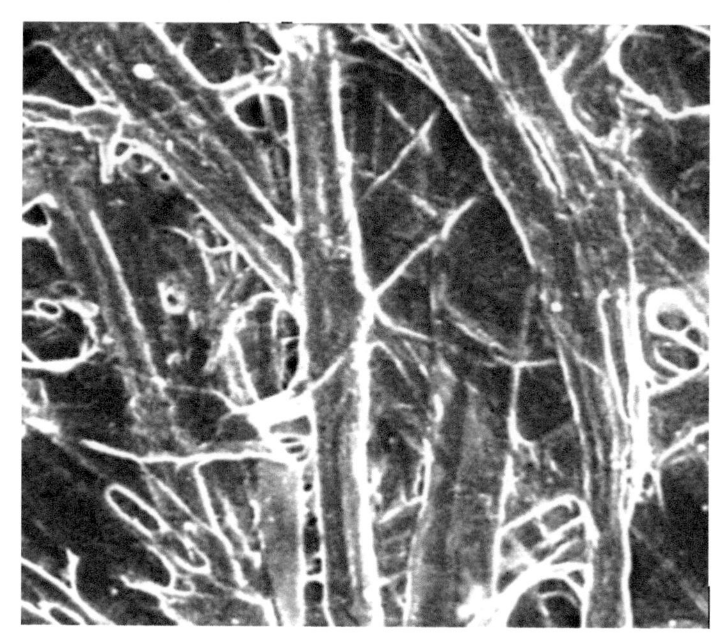

Figure 4.9 SEM of Kraft paper thermal stressed with NEO.

however, appears white. Samples' rough surfaces and displaced fiber orders are being indicated here. Deterioration of the samples is indicated by a shift in surface texture from smooth to rough. Extreme heat is the culprit [47], [69]. SEM image of Kraft paper aging in vegetable oil, displayed at ×200 magnification, is shown in Figure 4.9. It proves that band breakage does not occur when fibers are placed closely together. The fibers are densely packed when viewed under a high magnification. Because of this, thermal stress on solid insulating material is less than stress on Kraft paper soaked in mineral oil. New pressboard was viewed under a scanning electron microscope (SEM) at ×200 as shown in Figure 4.10. Both pictures show cellulose fibers in a fresh pressboard, which are aligned and attached to one another. There is also no band breaking. The SEM picture of mineral-oil-aged pressboard is displayed in Figure 4.11 at ×200 magnifications, respectively. The arrangement of the fibers in aged cellulose insulation has drastically shifted compared to that in new pressboard, revealing the layer of fibers in the middle. This points to the existence of bond disruption. It may be deduced that the pressboard suffers severe wear and tear due to the thermal aging process. SEM images of aged pressboard made from vegetable oil are shown in Figure 4.12 at ×200 magnifications, respectively. This finding suggests that the ordering of

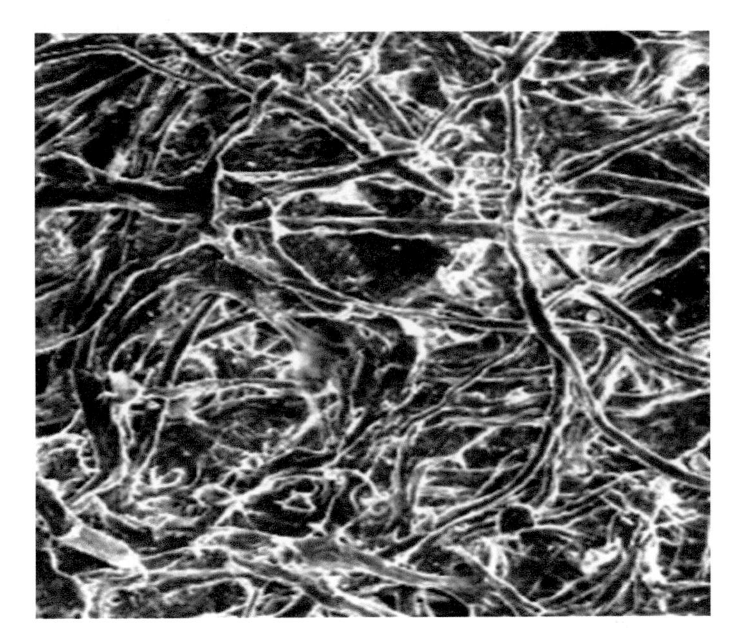

Figure 4.10 SEM of pressboard.

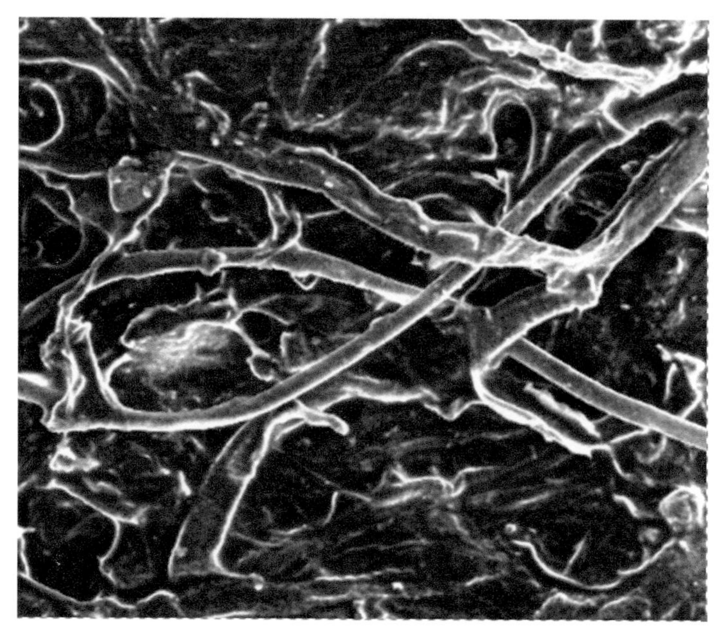

Figure 4.11 SEM of pressboard thermal stress with MO.

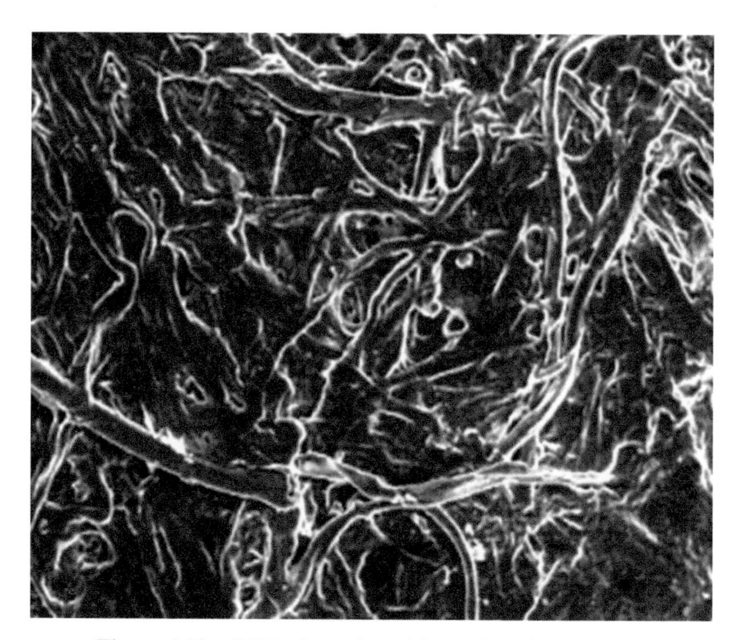

Figure 4.12 SEM of pressboard thermal stress with VGO.

cellulose fibers is not dramatically altered. Additionally, cellulosic fiber structural alterations are negligible. No bond breakdown is depicted in Figure 4.12. According to SEM research, vegetable oil slows down the deterioration of cellulose insulation compared to mineral oil.

4.4.3 FTIR measurements

The FTIR technique is used to take a comprehensive reading of all infrared wavelengths simultaneously. A solution was found with the help of a relatively straightforward optical instrument called an interferometer. It generated a novel form of signal that uniquely encodes all infrared wavelengths. Degradation products in oil samples can be analyzed quantitatively using this method [47]. Figure 4.13 depicts the FTIR spectrum of the oil sample analyzed. Multiple stretching and bending vibrations can be seen. Increasing or decreasing the connection length rhythmically via vibration is known as stretching. Changing the bond with a common atom or moving a group of atoms relative to the remainder of the molecule without moving the atoms in the group with respect to one another are both examples of bending vibrations [69].

Figure 4.13 shows the FTIR analysis of fresh mineral oil. Its spectrum analysis reveals a prominent absorbance peak at 2923.01 cm^{-1}, confirming the presence of alkanes via their distinctive C-H stretch. The presence of

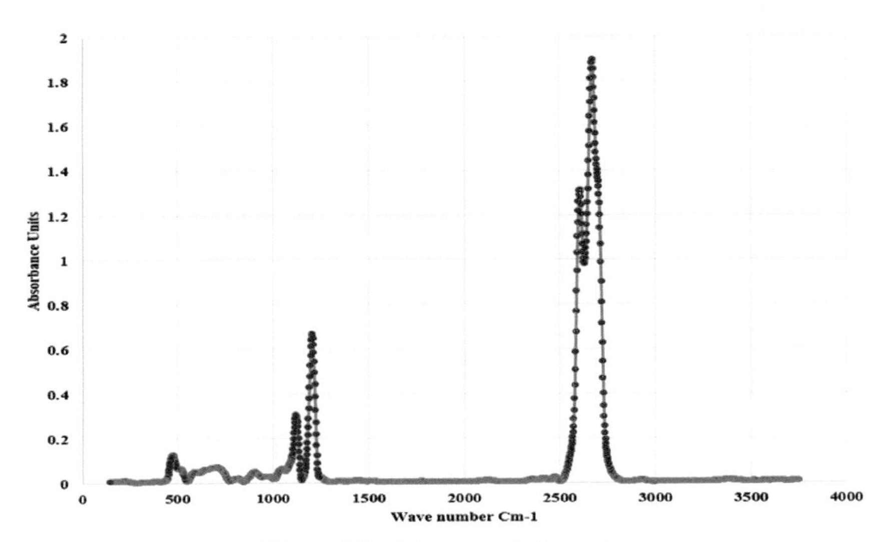

Figure 4.13 New mineral oil sample.

aromatic components in transformer oil is indicated by the appearance of a corresponding vibration at 1458.28 cm^{-1}, which is attributed to a C=C stretch. Next, an alkene signature, a faint absorption peak at 1402.165 cm^{-1}, is developed. Furthermore, an unexpected C-H bond happened around 760 cm^{-1}. As can be seen in Figure 4.14, however, the peak intensity of the absorbance curve for aged transformer oil is much lower than that of other oils. Aliphatic ketones are also present, as evidenced by the introduction of additional components between 1257.51 and 989.42 cm^{-1}. According to this, transformer oil will undergo a chemical change when subjected to high temperatures. Consequently, it caused the oil samples to become contaminated with aging byproducts [69], [78].

Ester oil's C-H stretching characteristic peak, at 2925.83 cm^{-1}, indicates the presence of the methyl group as shown in Figure 4.15; the C=O stretching characteristic peak, at 1745.47 cm^{-1}, indicates the presence of the carbonyl groups. In addition, a minor peak at 1458.09 cm^{-1} has been seen, which is C=O and is indicative of ketones. At 721.72 cm^{-1}, we see a last peak, which is caused by the superposition of the CH_2 rocking vibration and the out-of-plane vibration of cis-disubstituted olefins. See Figure 4.16 for an illustration of how the aging process weakens the absorption intensity of vegetable oil. Therefore, we concluded that the stability of vegetable oil is greater than that of mineral oil [69], [78].

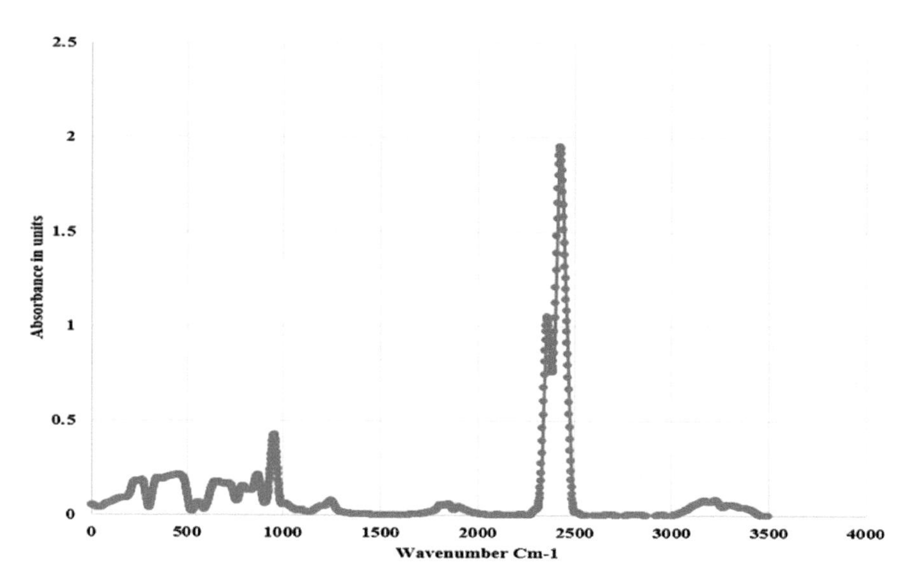

Figure 4.14 Thermally stressed mineral oil sample.

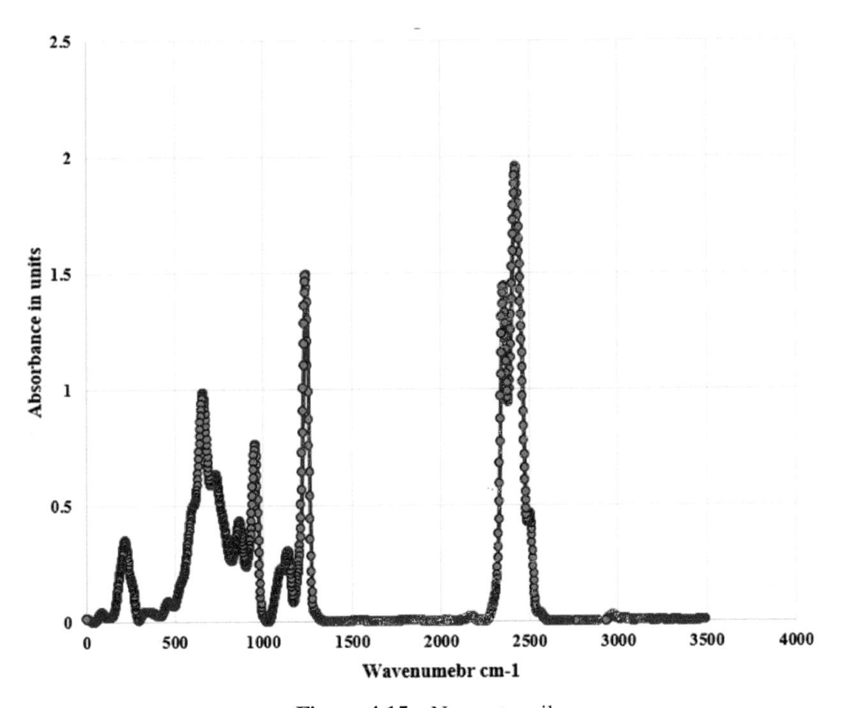

Figure 4.15 New ester oil.

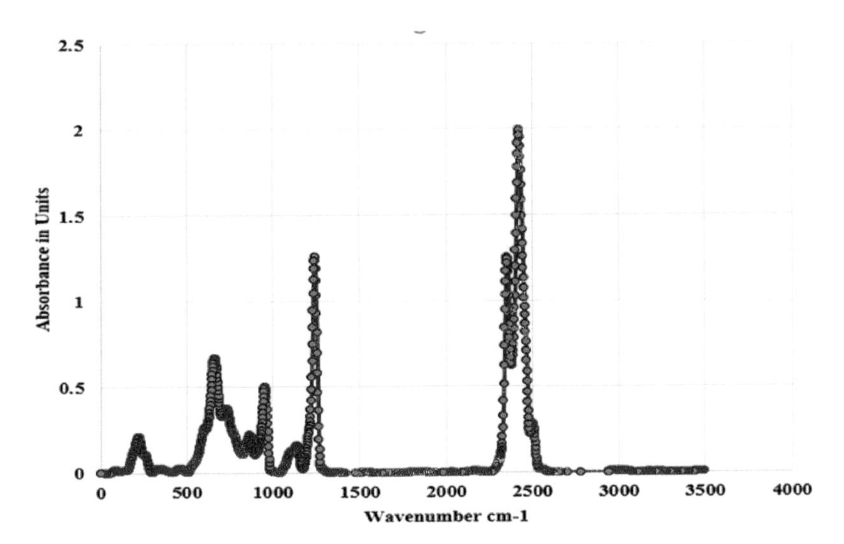

Figure 4.16 Thermally stressed ester oil.

4.4.4 Molecular spectroscopy

The monitoring and evaluation of green liquid dielectrics, especially in electrical equipment like transformers and circuit breakers, might benefit from Raman spectroscopy's usefulness and versatility as an analytical tool. The quality and integrity of green liquid dielectrics are crucial to the safe and dependable operation of electrical systems, where they play important roles as insulating and cooling fluids. Raman spectroscopy is very useful for determining the compounds present in green liquid dielectrics. When a dielectric is subjected to monochromatic laser light, the resulting Raman spectrum reveals the molecules' unique vibrational modes. The ability to ascertain the dielectric's chemical makeup is helpful for verifying its purity.

The presence of contaminants or degradation products in green liquid dielectrics necessitates constant monitoring. In order to identify and quantify these impurities, Raman spectroscopy provides a useful tool. Changes in the Raman spectrum can be used to detect various substances, such as water, gas bubbles, or solid particles. Because they can reduce the dielectric's effectiveness, identifying these impurities is critical to preventing equipment failure.

The degradation of green liquid dielectrics can be tracked with the help of Raman spectroscopy. The chemical composition of these fluids may shift as they age and are subjected to thermal and chemical stressors. Raman spectroscopy is useful for identifying structural changes or degradation products in molecules. This data is crucial for identifying the dielectric's overall condition and whether or not it needs to be repaired or replaced.

Raman spectroscopy can be used for quality control in the production of eco-friendly liquid dielectrics. Before incorporating dielectric fluids into electrical equipment, manufacturers can use this method to confirm that the dielectrics' chemical composition and purity fulfill required standards.

In-service monitoring of green liquid dielectrics in transformers and other electrical systems is possible with Raman spectroscopy. This allows for non-intrusive, real-time dielectric state monitoring while equipment is in use, allowing for immediate response to problems. Raman spectroscopy allows scientists to examine the properties of green liquid dielectrics at different temperatures and pressures. Research like this can improve electrical insulation by increasing its dielectric performance and durability.

5

Diagnosis Methods

5.1 Condition and Monitoring

Condition monitoring is the practice of checking in on a machine to see how it is doing and how far along it is. Since unanticipated breakdowns of machines and components lead to major accidents and enormous economic loss, this is preventing the breakdowns and extending the lifespan of machineries. Since transformers are one of the most important parts of a power system, condition monitoring of machinery is critical. Transformers can range in size from a few kVA to several hundred MVA. The replacement cost of a transformer might increase for a variety of reasons. The price to replace it might range from a few hundred to a million dollars, depending on its dimensions and voltage. The transformer has a 20–35 year design life. By practicing good upkeep, its lifespan can be extended to 60 years. Transformer insulation design is an important but often overlooked phenomenon. There is age-related variation in the resistance of transformer insulation [82], [83].

Transformers often utilize an oil-impregnated paper insulation method. The lifespan of the insulation was drastically cut short by the aging issue. Changes in insulating qualities that cannot be reversed constitute this phenomenon. Thermal aging, electrical aging, chemical aging, impulse aging, and environmental aging are just few of the processes that contribute to the aging process. Thermal aging is the principal cause of deterioration, and it causes insulation to undergo a variety of physical and chemical changes, including polymerization, depolymerization, and a shift in the fiber order of solid insulation. When insulation undergoes thermal aging, it either shrinks in size or expands over time. When a transformer is being tested prior to installation or during maintenance, an electrical phenomenon can occur. High impulsive voltage is delivered to the transformer during testing to determine how well it can handle the voltage without sustaining damage to its winding insulation or experiencing an early stage of failure. However, various circumstances during maintenance cause the operational current of a transformer to rise over what is listed on the nameplate [84], [85]. The temperature rises above the threshold level and the transformer fails before its time.

There are two main types of impulse phenomena, and these are switching impulses and lightning impulses. The abrupt opening or closure of the transformer, or the disconnection of a substantial inductive load, causes a switching impulse. However, a lightning impulse happens when lightning strikes the transformer directly. The insulation of a transformer is sensitive to both these factors and environmental variations [86], [87].

Normal and faulty transformer performance can be evaluated with the help of operational and accelerated testing. The accelerated test is a high-pressure experiment performed in a controlled environment. However, these conclusions are not applicable to a transformer in its typical operating environment, nor are they economically viable. Due to its excellent operating dependability under normal conditions, failure data of transformer is not achievable during regular test [88]. There are currently two methods for monitoring transformer conditioning in use: online and offline. Online monitoring and assessment of transformer performance under load conditions advises corrective action to be taken before the transformer reaches a critical state [89].

The CIGRÉ Working Group discovered that only 19% of failures were caused by the windings, while 41% were caused by OLTC. Mechanical causes account for 53% of failure onset, while dielectrics cause 31% of failure onset. The windings account for 26.6% of the total losses in a transformer without a tap changer, the magnetic circuit for 6.4%, the connection lead out for 33.3%, the tank and dielectric fluid for 17.4%, the other accessories for 11%, and the tap changer for 4.6%. According to Sahu's research, there is a 26 out of every 1000 chance that a transformer may fail. The aforesaid analysis leads us to the following conclusion [90]: the life of a transformer is same as the life of its insulation.

The transformer breakdown paten can be determined using the bath tub curve. These curves can be broken down into three sections, the first of which is caused by infant mortality. The second half of the curve represents a consistent failure rate, while the final segment represents a cure owing to advanced age [91], [92].

According to research by Grechko et al., 51% of transformer failures occurred during maintenance in the first five years of the transformer's life. Damage to the winding or the decomposition of the winding as a result of short circuit forces, and damage to the transformer's bushing [201] are the most common reasons for transformer failure.

Measurements of ratio, winding resistance, short-circuit impedance and loss, excitation impedance, loss dissipation factor, and capacitance are all part of a standard transformer condition test that helps technicians locate issues like winding faults, wind deformation, joint failure, dielectric breakdown, and oil moisture and contamination. In addition, a unique test called partial discharge detection is performed to quickly locate an internal problem in the transformer's winding. Winding movement and lost conductor can be detected using frequency response analysis. The temperature of hot spots on the transformer's exterior can be detected with an infrared image, and the contamination rate of the insulation can be determined by measuring its degree of polymerization [93], [94].

5.2 Methods for Tracking Transformer Health Status

5.2.1 Online methods for detecting partial discharge

Transformers experience partial discharge during the energizing process because of flaws in the insulating design. There is no electric current established between the electrodes during partial discharges. However, if the amplitude of the discharge is large enough, it can cause the insulation of the electrical equipment to fail. Multiple techniques can be used to detect a partial discharge: electrical, chemical, acoustic, and ultrahigh-frequency (UHF). Detecting the auditory sound and measuring the resulting electrical signal is the standard procedure for PD. Laboratory chemical analysis of oil is also used to detect PD. The PD acoustic technique makes use of an appropriate signal generating device to produce PD pulse signals. Acoustic emission sensors installed on the interior wall of the transformer tank or in the oil itself pick up signals sent to the transformer. In the beginning, less sophisticated piezoelectric sensors are utilized instead of optical ones. Fiber optic sensors are often utilized in niche applications. The PD permissible limit varied from 100 PC up to 500 PC depending on the transformer's size and voltage rating [95]–[97].

5.2.2 Frequency response analysis

Overloading a transformer causes the windings to shift because of the tremendous electromechanical force created. This necessitates early identification of winding motion. No other conditioning or monitoring technology can detect the winding motion. In terms of fingerprint analysis, frequency response analysis is the method of choice for identifying winding motion. In the past, the winding movement was detected using a low voltage impulse approach. This technique involves applying a small impulse voltage to one end of the coil and then measuring the resulting response at the other end. The signal's lifespan is cut off abruptly at both ends. This technique requires a short amount of time to diagnose; however, the findings are impacted by the method's restrictions on frequency range and impulse form. The test leads degrade with repeated readings. Nowadays, winding displacement can be detected using SFRA. This technique employs a large measurement frequency band. This method's benefits include a high signal-to-noise ratio, a wide frequency range, and high frequency resolution [87].

The above response exemplifies normal operation (Figure 5.1). The response is simulated without fault. In addition to mutual inductive and capacitive coupling between the winding elements, the characteristics of FRA responses of windings, such as multiple resonances and antiresonances, are effectively determined by the large number of inductance and capacitance elements present in the equivalent circuit of the transformer winding over a broad frequency range. At 399.139 kHz, the first resonance is being experienced. Once the resonance frequency is passed, the transformer winding's inductance takes over. When the winding's magnetic effect tries to grow after the first resonance point, the inductance screens it out. This occurs again,

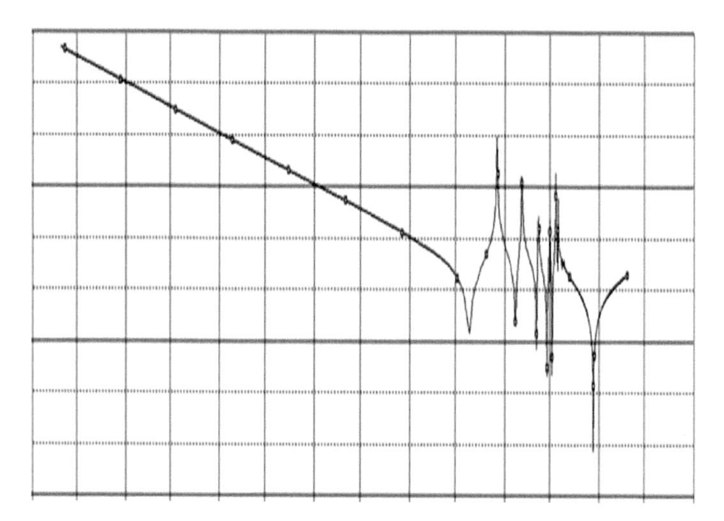

Figure 5.1 Finger print transformer SFRA.

increasing the number of resonance spots in the medium frequency range. The series and shunt capacitance of the windings totally cancel out the influence of the winding inductance after the medium frequency range. The wave form measurements indicate a current of 451.645 A. The first resonance point might be used as a starting point for further analysis.

The fault state between turns is being simulated here. In Figure 5.2, the fault is introduced artificially in the transformer's fourth turn of windings. It is seen from comparing the wave forms with and without a flaw, as shown in Figure 5.2. At 429.068 kHz, we find the initial point of resonance. There is also a small shift in the wave form at medium frequencies compared to the wave form under normal conditions. Large frequency shifts, however, occur between 2 and 4.2378 MHz. Inter-turn fault current is found to be 500.925 A. The current will rise to 49.28 A from the no-fault state. Transformer insulation will be compromised and windings will burn out as a result of the increased current. This model is simulating a fault condition between two turns. In Figure 5.3, the fault is deliberately introduced between the third and fourth turns of the transformer's windings. It is seen from comparing the wave forms with and without a flaw, as shown in Figure 5.3. We also draw parallels between this wave form and a fault that we have already covered. At 427.471 kHz, we see the appearance of the first resonance point. The observed wave form is significantly off-center from the reference set for turn-to-turn fault

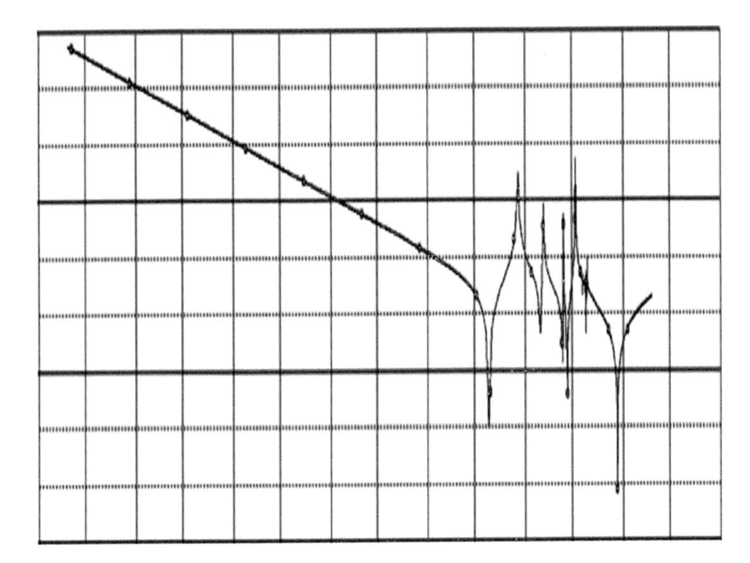

Figure 5.2 SFRA with inter-turn fault.

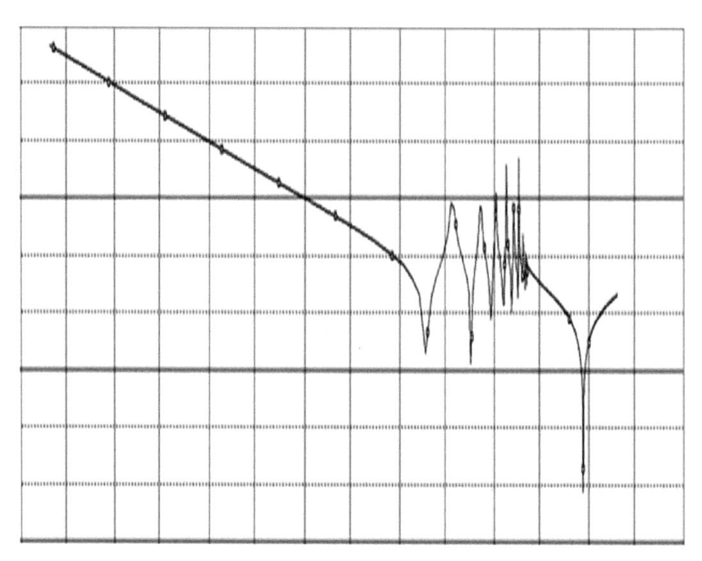

Figure 5.3 Turn-to-turn fault.

conditions, shifting from 1.4417 to 4.3697 MHz. Inter-turn fault conditions have a current of 562.264 A measured, with an increase of around 110.619 A compared to no-fault conditions. The insulation failed because of the sudden increase in temperature.

At the time that PD happened on the transformer winding, the aforementioned wave shape was captured. Figure 5.4 fundamental wave form has been strayed from. In contrast to the wave form displacement observed at turn faults, the resultant wave shape is completely unique. At 409.600 KHz, we see the appearance of the first resonance point. There is a significant shift in the wave shape between 409.600 KHz and 4.4266 MHz, where the PD on the transformer winding occurred.

When an electrical impulse is supplied to the supply end of a transformer's winding, the resulting wave form looks like this (Figure 5.5). There is a departure in wave form from the norm. The resulting wave form deviation is compared to a standard. Displacement of the wave form is obviously distinct from the winding fault that has been previously mentioned. Such a shift will take place only at the precise moment, a lightning impulse strikes. At 409.600 KHz, we see the appearance of the first resonance point. At the initial resonance point about 20 MHz, the wave shape shifts as shown in the figure above. The insulation of the transformer will be severely compromised as a result.

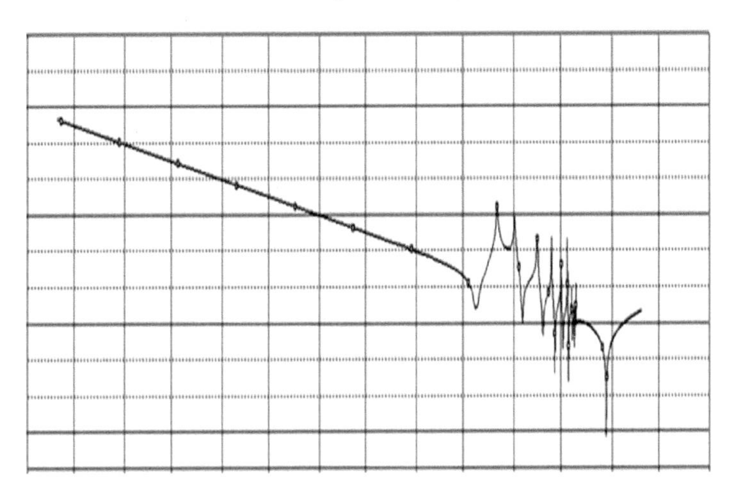

Figure 5.4 SFRA under partial discharge fault.

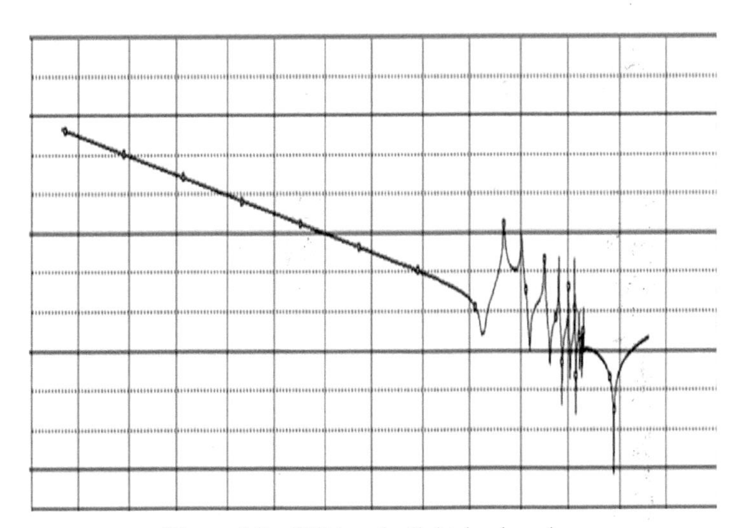

Figure 5.5 SFRA under lightning impulse.

5.2.3 Technique for restoring voltage

Type of paper, pulp composition, thermal upgrading, moisture content, and operating temperature all play a role in the pace of paper degradation. The rate of a transformer's deterioration is particularly sensitive to its moisture level. This technique is used to determine the health of oil. The transformer's insulation is made of paper. It is also employed in the assessment of insulation's moisture content. The first step is a brief application of the source voltage. The circuit is then cut off from its power supply and briefly shorted out. Charging takes longer than the short circuit time. The return voltage at the electrode is then measured when the short circuit is broken. Maximum return voltage, period at which maximum peak is obtained, and starting slope of the return voltage are taken into account when determining the state of oil and paper. The transformer must be unplugged from the power source for this method to work [98], [99].

5.2.4 Internal defect detector

The in-built malfunction sensor is a mechanical detector. The transformer's internal failure is the only trigger for its activation. It constantly monitors the sharp increase in air pressure above the oil. Internal fault is the fundamental reason for the quick increase in oil pressure. Because of this, the transformer's operating temperature rises, causing the oil to evaporate. The sudden increase in pressure causes the spring to unwind. The tank's internal fault detector may be seen at its other end from the outside [100].

5.2.5 Temperature of the internal winding

Fiber optic sensors are used in this technique to determine the temperature of the windings. The fiber optic cable is installed along the winding of the transformer during production. This means that it can keep checking the transformer's temperature at all levels. However, not only would this approach waste money, but mechanical stress will also be generated from the fire itself. The importance of fiber cannot be overstated. Hot spot temperature is another area where fiber optic point sensors come in use. These sensors are embedded within the target zone during production. This approach, however, is equally impractical [101], [102].

5.2.6 Tracking the tap-switch status

Tap changer contact deterioration can be measured with this test. The tap changer is equipped with a motor that takes current from the source, and this current is constantly monitored and its "fingerprint" current wave checked often. There must be a problem with the tap changer if the fingerprint is stated to have changed [103], [104].

5.2.7 Analysis of the furan component

Insulation made from cellulosic materials is derived from a glucose polymer that is 90% cellulose, 7% hemicellulose, and 4% lignin. Inside the transformer, the polymer chain is deteriorating owing to electrical, thermal, and chemical stress. Degradation of solid insulation produces copious amounts of carbon monoxide, carbon dioxide, organic acids, and free glucose. The breakdown of these molecules results in the formation of furans. Furanic compounds include 2-furfuraldehyde, 2-acetylfuran, 5-methyl-2-furfuraldehyde, etc. Very little amounts of these are made in transformer oil. After removing a sample of oil from the transformer, it is examined using HPLC. Analysis of the paper's furan concentration reveals information on the paper's condition and polymerization level. Increases in furanic components are a good indicator of how quickly paper is aging. In normal operation, the new paper insulating transformer releases 1.7 ng/g of paper every hour of furan. An additional half milligrams of furfural per gram of paper is produced when transformer breakdown is accelerated. However, a transformer's service life is estimated to be 20–30 years [105]–[107].

5.2.8 Expert system and diagnostic program

In contrast to more traditional methods of analysis, it produces reliable findings. When compared to traditional analysis, the results from this approach are more trustworthy. The software takes in information about the gas, moisture, and dielectric strength levels of the test samples and draws conclusions about the transformer's health. The transformer parameters will trigger the alert when they go above the threshold value, a need of all expert systems. Recently, an intelligent system [215], [216] has been employed to diagnose transform errors.

6

Thermography

6.1 Real-time Case Studies

Equipment emits infrared radiation when its temperature rises above absolute zero. Since this radiation gives out thermal energy, it is invisible to the naked eye. Infrared thermography [108] is used to create a visible thermal image from these otherwise unseen heat radiations. The resistivity of each piece of equipment varies in service because of the differences in the materials used in their construction. Heat is produced by these devices when an electric current is run through them, and this heat is proportional to the square of the current and the resistance. Consequently, more resistance results in more heat being generated. Electrical components deteriorate with age, which raises their resistance and causes the equipment to overheat. The result is premature equipment breakdown and consequential financial loss. It has negative effects on the economy, the environment, and the health of living things. As a result, it is crucial to keep a close eye on the temperatures at which various electrical components are working at all times. Traditionally, equipment temperature monitoring has relied on thermocouples and resistance temperature monitors. It keeps an eye on the temperature all the time, and if it gets too high, it sets off an alarm, trips, etc. However, the thermal image of an item is not provided by these. Modern condition monitoring of electrical equipment makes use of infrared thermal imaging cameras. Infrared thermal imaging device generated thermal profiles of live electrical systems with no downtime. Both a thermal image and a temperature scale are included in these thermal profiles. This appears in a distinct hue, denoting the specific piece of machinery at play [109]. Here, a thermal imager is used to keep tabs on the status of the transformer.

As shown in the above case studies from Figures 6.1–6.5, by using a thermal imaging camera, we can easily find the hot spot of the transform. Suppose the hotspot temperature is above the threshold value, which infers that the transformer condition is critical. In addition, the thermal imager has the option to identify the thermally stressed area in an accurate manner.

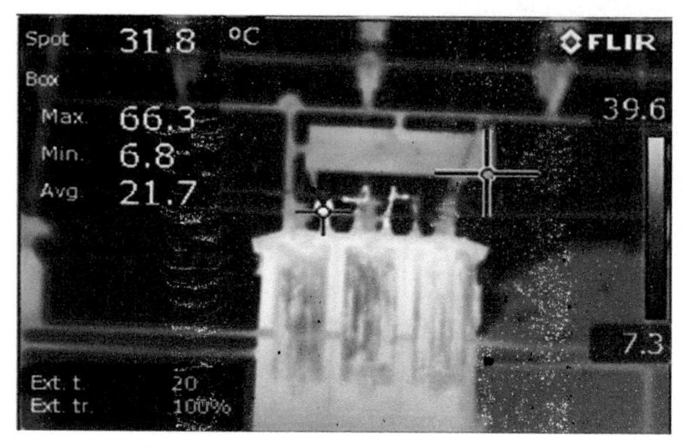

Figure 6.1 Thermal image of DT1.

Figure 6.2 Thermal image of DT2.

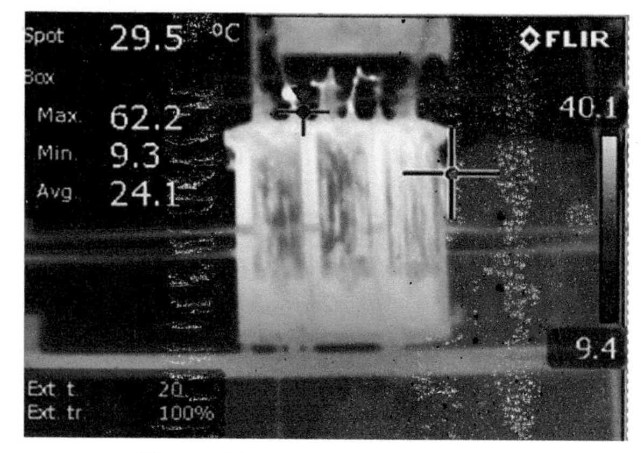

Figure 6.3 Thermal image of DT3.

Figure 6.4 Thermal image of DT4.

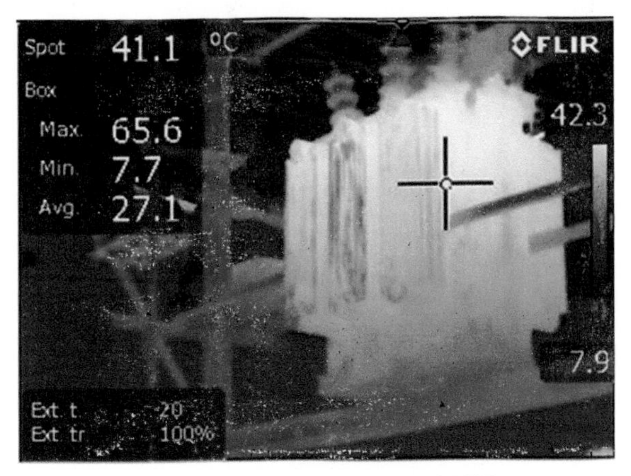

Figure 6.5 Thermal image of DT5.

Figure 6.6 shows a thermal image of a three-phase transformer bushing. The hot spot temperature is appearing on bushing in the range of 61.6 °C. These are due to aging byproducts such as moisture and dust deposited on the bushing; hence, proper cleaning is required. Also, check the oil level within the bushing.

The hottest part of the transformer is illustrated quite clearly in Figure 6.6. This was the point where the line branched off. The temperature range is at or above 150 °C. A yellowish-mixed-green color can be seen across the rest of the area, apart from the hot spot. The fact that other parts of the transformer are blue gives one the impression that its overall condition is satisfactory. However, the location of the place is becoming increasingly crimson. It manifests itself on the conductor joint as a result of sparking, which may be the result of a loosening of the line joint or corona. Because of the extremely precarious nature of this situation, immediate action is essential. The hottest region on the transformer is depicted quite clearly in Figure 6.7. This is the point at which the line will lead out. The temperature range at the site is 327 °C, which could be because the line or the corona is getting thinner. In addition, the significant spots of the transformer start to turn a reddish color, which is why it is essential to inspect the quality

Figure 6.6 Thermal profile of a transformer.

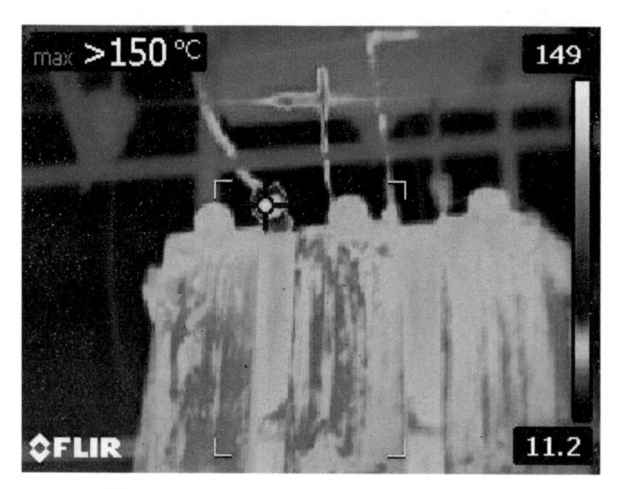

Figure 6.7 Hot spot appears on line joint.

and load profile of the transformer. The thermal image of a transformer that is 20 years old shows that the primary components of the transformer are reddish in color, as can be seen in Figure 6.8. The highest temperature ever recorded there was 124 °C. The spot where the sparks are coming from is still sparking, and this is because the point where the lines intersect has deteriorated. It can be deduced that the oil's quality is poor; hence, changing the transformer's oil is something that should be done as soon as possible. It can be seen in Figures 6.9 and 6.10 that the big areas are starting to appear on top of the transformer tank. In addition, the bushing of the transformer is in an extremely hot position; as a result, it is imperative that the level of oil in the transformer tank be checked.

Figure 6.8 Hot spot on line joint and transformer.

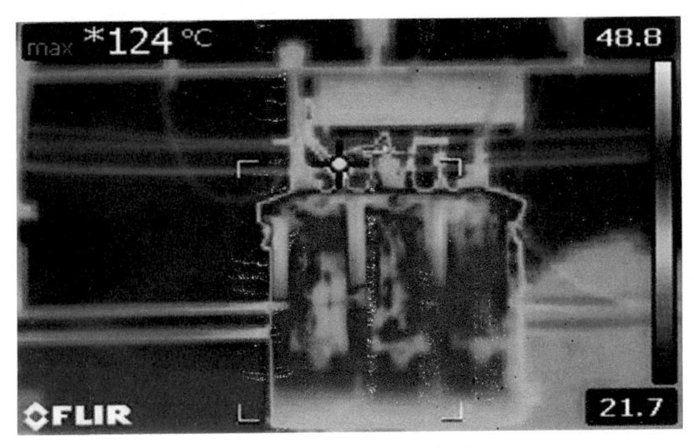

Figure 6.9 Transformer in danger.

Figure 6.10 Hot spots appear on top of the transformer tank.

7

AI Technique

7.1 Artificial Intelligence for Transformers

The utilization of artificial intelligence (AI) in condition monitoring and predictive maintenance for transformers is of utmost importance in upholding the stability and dependability of power systems. The comprehensive assessment of a transformer's condition can be achieved through the collection and integration of data from several sensors, including temperature sensors, moisture sensors, and dissolved gas analysis (DGA) sensors, by artificial intelligence (AI) systems. These systems employ advanced anomaly detection techniques to discover any deviations from typical operational parameters by extracting relevant information from the data. This analysis facilitates the prompt identification of potential errors or irregularities by AI, hence enabling timely intervention and the implementation of preventive actions. Furthermore, the utilization of artificial intelligence (AI) facilitates the prognostication of the remaining useful lifespan of transformers through the examination of past data and the identification of patterns in their functioning. The utilization of these forecasting capabilities enables maintenance teams to adopt a proactive approach in planning and organizing maintenance tasks, resulting in a substantial reduction in the likelihood of unforeseen failures and a minimization of operational downtime. In addition, AI-powered health indices are calculated in order to offer a full evaluation of the transformer's overall condition. These indices play a crucial role in the prioritization of maintenance activities, enabling a focused strategy that concentrates on key matters and optimizes the allocation of resources. AI systems also provide prescriptive maintenance advice by conducting thorough data analysis. This empowers maintenance professionals to proactively address prospective concerns. The aforementioned recommendations provide guidance for the implementation of particular diagnostic tests, the substitution of components, or the arrangement of necessary maintenance activities. The integration of real-time monitoring and alarm systems into artificial intelligence technology enables prompt notifications to be generated upon the detection of key changes or probable breakdowns inside the transformer. The aforementioned alarms provide rapid responses and essential interventions, thereby minimizing disruptions to the power supply and mitigating potential harm to both the transformer and the broader power system. By utilizing artificial intelligence (AI) for the purpose of condition monitoring and predictive maintenance, power system operators have the potential to greatly enhance the operating efficiency and reliability of transformers. The capacity to proactively identify and mitigate possible challenges prior to their escalation contributes to the reduction of operating expenses and the improvement of overall system efficacy. By leveraging AI-powered analysis and suggestions, maintenance procedures may be enhanced and shortened, resulting in enhanced durability and robustness of transformer assets within the power grid [111]–[113].

7.2 Fault Diagnosis and Troubleshooting

The use of AI has revolutionized the procedures of fault diagnosis and troubleshooting for transformers, both of which are essential to assuring the dependable running of power systems. Transformer failures and irregularities can be discovered and fixed by power system operators with the help of AI algorithms. By analyzing complicated data patterns and connections, AI systems enable rapid and precise diagnosis of problems and their causes. By conducting such a thorough investigation, AI helps technicians gain a better comprehension of the components that contribute to problems, which speeds up the troubleshooting process and allows for more targeted interventions. In addition, operators can analyze the severity and potential ramifications of identified defects thanks to AI-based fault diagnosis. Artificial intelligence (AI) helps power grid operators prioritize and allocate resources effectively by offering actionable insights into the type and scope of the difficulties, so that they may fix the most urgent failures first and reduce the system-wide impact as little as possible. Artificial intelligence systems also help create sophisticated diagnostic models that can mimic different fault scenarios, which is useful for foreseeing issues and taking preventative measures to avoid disruptions. Improved operational efficiency and grid resilience are two outcomes of artificial intelligence's adoption in transformer problem identification and troubleshooting. Artificial intelligence technologies drastically cut downtime and maintenance costs by streamlining the diagnostic process and speeding the discovery of defects. The entire stability and dependability of the power system is improved by the ability to proactively resolve any issues before they escalate, guaranteeing a constant and reliable supply of electricity [115]–[117].

7.3 Energy Efficiency and Loss Reduction

Integrating AI in transformer applications is crucial for increasing energy efficiency and loss reduction, two crucial goals for power systems. In order to find ways to increase efficiency and decrease losses in transformer operations, AI-enabled solutions can provide advanced analytics and real-time monitoring. In order to maximize energy efficiency and reduce waste, AI algorithms can analyze both historical and real-time data to find patterns and inefficiencies and then execute focused tactics to do so. Better load management and distribution planning are possible with the use of AI-based predictive models that project future energy consumption patterns. As a result of AI systems' improved ability to foresee demand changes, energy resources can be better allocated, and transformer operations can be optimized, keeping transformers running within their most efficient range. This preventative measure improves the power system's overall energy efficiency by decreasing energy waste.

Also, AI systems can find and assess the root causes of energy waste, such as poor voltage regulation, overloading, or improperly configured transformers. Artificial intelligence (AI) can help reduce energy waste by keeping tabs on operational parameters and system performance. This preventative measure guarantees the transformers are working at peak performance and helps reduce wasteful energy loss. Monitoring and managing characteristics like voltage levels and harmonic distortions, AI-enabled devices can improve power quality. Artificial intelligence (AI) aids in reducing energy waste caused by voltage variations and other system anomalies by providing steady and constant power delivery. By lowering the stress and wear caused by unstable power inputs, this method not only improves the power system's overall efficiency but also increases the lifespan of transformers. Power utilities can save a lot of money, lessen their impact on the environment, and lengthen the lifespan of the power grid by employing AI to enhance energy efficiency and decrease loss in transformer applications. Power systems can satisfy increasing energy demands with less waste and fewer negative environmental effects thanks to the continual monitoring and optimization made possible by AI systems.

7.4 Optimal Operation and Control

Maintaining power system efficiency and reliability requires optimal transformer operation and control. Transformers can function at their peak with the help of artificial intelligence, which allows for the implementation of sophisticated control schemes and real-time optimization approaches. To ensure that transformers are operating at peak efficiency, AI-based systems continuously analyze data from a wide variety of sensors, past records, and grid characteristics. Also, AI helps with load forecasting, voltage control, and power flow optimization, which allows for better management of power distribution and keeps transformers running at peak efficiency. Artificial intelligence (AI) systems can monitor and respond to grid variables such as load demand, voltage variations, and other factors to provide consistent and reliable electricity delivery at all times. These features improve the transformers' operational efficiency and help keep the electricity grid stable and reliable.

In addition, operational parameters can be automatically adjusted in response to dynamic changes in the grid because of adaptive control mechanisms made possible by AI-driven control techniques. Artificial intelligence (AI) systems can proactively regulate load distribution, reduce voltage fluctuations, and optimize power transfer by constantly monitoring grid conditions and transformer performance, thereby reducing energy losses and making better use of transformer assets. Power utilities may improve energy efficiency, decrease operating costs, and boost the performance and stability of the power grid by utilizing AI for efficient operation and management of transformers. Transformer asset utilization is improved thanks to the incorporation of AI-driven control techniques, which also aid in the efficient administration of the power network and guarantee a consistent and reliable supply of electricity to customers.

7.5 Life Cycle Assessment

Integrating AI technology into life cycle assessment (LCA) and asset management can greatly improve these practices, which are essential for assuring the long-term performance and sustainability of transformers. Power utilities may analyze transformers' long-term performance and evaluate their environmental impact across their entire life cycle by utilizing AI-driven analytics and data processing capabilities. Throughout a transformer's service life, critical performance parameters including energy efficiency, operational dependability, and environmental effect can be monitored in real time by AI-enabled systems. In order to help operators make educated decisions on maintenance schedules, refurbishments, or replacements, AI algorithms may analyze historical data and real-time operational parameters to get useful insights into the overall health and performance of transformers [120].

Furthermore, asset management systems powered by AI can improve transformer maintenance and replacement planning and resource allocation. The remaining useful life of transformers can be predicted by AI systems by combining data from LCA assessments, historical performance records, and real-time monitoring data. With these foresightful abilities, utilities may better manage their transformer assets, maximizing utilization of resources and reducing the likelihood of breakdowns. Further, maintenance tasks can be arranged according to the real condition and performance of transformers with the help of AI-driven asset management systems, rather than at specified intervals in advance. AI technologies aid in optimizing maintenance operations, decreasing downtime, and extending the total life cycle of transformers by prioritizing maintenance efforts based on the criticality of assets and the severity of discovered issues. Transform your power company's operational efficiency, maintenance expenses, and environmental impact with the use of artificial intelligence by implementing a life cycle assessment and asset management practice for transformers. Transformers can be reliably and sustainably operated over their entire service lives thanks to AI systems' ability to continuously evaluate and predict their condition [121], [122].

7.6 Data-driven Decision Making

Decisions in the current management of transformers are increasingly being driven by data, which is crucial for improving transformer efficiency and maintaining reliable power networks. Uninterrupted operation of transformers is crucial since they constitute the backbone of power distribution networks. Utilities and organizations may improve transformer health and grid dependability with the help of data analytics, machine learning, and real-time monitoring. Predictive maintenance is a crucial part of data-driven decision making for transformers. In order to foresee when maintenance will be required, this method makes use of both historical and real-time data. Data patterns can be analyzed by machine learning algorithms, allowing for preemptive repairs and better problem prediction. A more effective grid is the outcome of this method's reduced downtime and operational expenses.

Fault detection is an important part of data-driven decision making. Dissolved gas analysis (DGA) and oil quality are two examples of crucial characteristics that may be monitored continuously to detect anomalies and potential defects as soon as they occur. These issue categories can be categorized by machine learning models, leading to more precise maintenance and repair. With this method, transformers last longer, function better, and experience fewer power outages [123].

One more benefit of data-driven decision making is effective resource allocation. Utilities can prioritize the maintenance of transformers by analyzing data to determine which ones require it the most. Manpower efficiency is increased, and operating expenses are decreased, thanks to this optimized resource management that makes better use of available resources. Data-driven decision making also includes studying the effects on the environment and the weather. To learn how weather and environmental conditions have changed throughout time, it is necessary to examine historical data. Utilities can better respond to changing conditions and reduce risks with this information in hand.

7.7 Assessment of Transformer Health

The concept of perpetual learning and adaptation has greatly advanced the field of transformer health assessment, enabling machine learning models to continuously improve and adapt to new data and evolving conditions. Transformers play a crucial role in power distribution networks, as their proper functioning is important for ensuring the stability and dependability of the electrical grid. Traditional assessment methods often struggle to keep pace with the dynamic nature of transformer operation, making continual learning and adaptation an invaluable approach. The practice of continual learning is essential in maintaining the currency of machine learning models by incorporating the most recent data. Transformers are subject to dynamic environmental circumstances, fluctuating loads, and altering usage patterns. Machine learning models have the capability to promptly detect developing trends, abnormalities, and possible issues by continuously assimilating fresh data. The capacity to adapt enables preventative maintenance, thereby mitigating the likelihood of abrupt malfunctions and enhancing the overall lifespan of transformers [124].

Adaptive models provide exceptional proficiency in effectively managing the intricate and interrelated characteristics of transformer health evaluation. Various data sources can be analyzed, including operational data, environmental variables, sensor readings, and historical records. These models possess the ability to adjust to variations in data patterns and autonomously revise their evaluation criteria in order to accommodate alterations in transformer behavior. The ability to adapt is of utmost importance when it comes to identifying subtle or slow changes that may go unnoticed by conventional assessment methods. In addition, the ongoing process of learning and adaptation plays a crucial role in enabling the implementation of predictive maintenance strategies. Through the examination of trends and patterns over a period of time, machine learning models has the capability to predict the timing at which a transformer may want repair or replacement. This enables utilities to effectively strategize and allocate resources in a manner that optimizes efficiency. The implementation of this proactive method yields substantial reductions in downtime, operational expenses, and the likelihood of unforeseen transformer failures [125], [126].

7.8 Deep Learning

In terms of power grid management, the application of deep learning methods to the evaluation of transformer health problems is a major step forward. Power transformers play a crucial role in the reliability of the electrical grid and must operate continuously. Deep learning, a type of machine learning, provides engineers and utilities with a powerful framework for analyzing large datasets and gaining insights regarding transformers' operational state.

Deep learning is widely used for data-driven fault detection in transformer health monitoring. Sensor data collected by transformers can be analyzed by deep neural networks due to their capacity to interpret complicated data patterns. Deep learning models can detect anomalies in parameters such as temperature, load, voltage, and others that are used to diagnose problems. By allowing for early diagnosis and preventive repair, this feature increases transformer dependability and decreases the likelihood of expensive outages.

Predictive upkeep is another advantage of deep learning models. These algorithms can predict when a transformer will need repair or replacement by continuously learning and adapting to new data as it becomes available. By helping utilities better schedule maintenance efforts, decrease downtime, and increase the lifespan of important transformer assets, predictive maintenance can result in substantial cost savings.

In addition, deep learning helps improve processes like data processing and decision making. Intricate linkages and interrelated variables are common features of complicated power distribution systems. Engineers and utilities can benefit greatly from the insights uncovered by deep learning models when they apply them to data including information on transformers.

7.9 Remote Monitoring and Management

Transformer insulation management and remote monitoring are critical components of today's electrical grid. Insulation is critical for the continued reliability of transformers, which play a major role in power distribution networks. In order to optimize transformer performance, remote monitoring is used to continuously evaluate insulation health through the use of cutting-edge technologies, sensors, and data analysis.

Collecting real-time data from multiple sensors installed within transformers is a major advantage of remote monitoring. These sensors provide a holistic picture of insulation conditions by measuring things like temperature, load, gas concentrations, and oil quality. Engineers and technicians can obtain this information in real time thanks to remote monitoring devices that send it to command centers.

Automatic alerts and notifications are a part of remote monitoring. The monitoring system will send out alerts if it finds any irregularities or potential insulation issues. This quick action reduces the likelihood of expensive transformer failures and outages.

Condition-based maintenance is another benefit of remote monitoring. Utilities no longer have to rely on predetermined schedules for maintenance and refurbishing of transformers thanks to ongoing insulation health assessments. This method improves resource allocation while decreasing maintenance expenses and downtime.

7.10 Continual Learning and Adaptation

An innovative strategy for guaranteeing the safety and durability of transformers in power grids is to implement a system of ongoing learning and adaptation. The insulation of transformers is a crucial component in ensuring the reliability of the power grid. To maintain transformers in peak operating condition, to adapt to shifting operational situations, and to forecast and prevent prospective difficulties, continuous learning makes use of data analytics, machine learning, and real-time monitoring.

The ability to adjust to new information is a cornerstone of lifelong education. There are many different types of loads, environmental considerations, and operating circumstances that transformers may face. To make sure their evaluations of transformers are always up-to-date and accurate, continuous learning models are constantly updating their understanding of transformer behavior based on fresh data.

Because evaluating transformer insulation is such a complex and linked process, adaptive models perform exceptionally well in this role. They are capable of analyzing information gathered from many sensors (DGA, temperature, and load) and adjusting to new data patterns. To identify gradual or subtle changes in insulation health, this flexibility is crucial.

Constant improvement also manifests itself in features like predictive upkeep. Models can analyze past and current data to foresee when a transformer will need servicing or replacement. As foreseeable failures are avoided and downtime is minimized, the cost-effectiveness of this strategy becomes clear.

7.11 Life Cycle Estimation

To guarantee the continued efficiency and dependability of transformers in electrical distribution networks, it is essential to accurately predict how long their insulation will last. Insulation on a transformer is vulnerable to wear and tear from a variety of sources, such as temperature swings, changes in load, and the elements. By providing a data-driven and preventative approach to maintenance and management, machine learning is a valuable tool for predicting and extending the life cycle of transformer insulation.

First, information is obtained, both historically and in real time, about the insulation of the transformer and its operational environment. The dataset contains information on a variety of parameters, such as temperature, oil quality, dissolved gas analysis (DGA), and prior maintenance logs. Identifying important characteristics or variables that affect insulation life cycle requires feature engineering. This ensures that the machine learning model can accurately capture the connections between these variables and the insulation's remaining life expectancy.

Next, the data must be cleaned, missing values filled in, outliers removed, and the dataset formatted in a way that makes it easy to analyze. The designated label is used to indicate how much time is left in the insulation's useful life cycle; this information is typically gleaned from replacement or renovation records from the past. Which machine learning model to use is situational and data-specific. The likelihood of insulation failure can be estimated using a survival analysis model or a regression model for the remaining useful life.

The data collection must be split into training, validation, and test sets before any model can be developed. To evaluate the model's efficacy, it is first put through its paces on the validation dataset after it has been trained on the training dataset to understand the associations between the input attributes and the desired label. Mean absolute error (MAE), mean squared error (MSE), and the concordance index (C-index) are common evaluation metrics used to measure model correctness in survival analysis.

The remaining lifespan of transformer insulation can be predicted using machine learning models. Insights like this allow utilities and maintenance teams to plan ahead for maintenance and refurbishment, which in turn increases the lifespan of transformers and decreases the likelihood of unplanned outages. Keeping the model up-to-date with fresh data is crucial to ensure its flexibility to adapt to shifting operational conditions and developing insulating materials. In conclusion, estimating the lifetime of transformer insulation using machine learning is a cutting-edge strategy that improves the efficacy and dependability of power grids. It allows for more informed decision making, reduces maintenance needs, and improves transformer longevity [126]–[128].

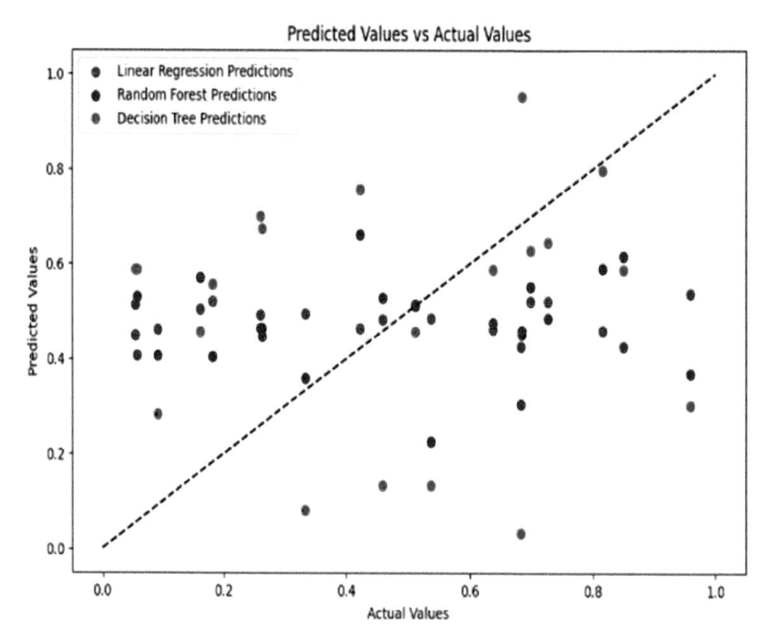

Figure 7.1 Comparative analysis on ML.

Results: Linear regression mean squared error: 0.07418987184028222

Random forest mean squared error: 0.08715167334307906

Decision tree mean squared error: 0.13271554286400317

The comparative analysis between linear, random forest, and decision tree algorithm for predicting the status of the transformer is shown in Figure 7.1 Lower values of mean squared error indicate better performance, as they represent a closer fit of the model's predicted values to the actual values. In this case, the linear regression model seems to have the lowest MSE, suggesting that it performs slightly better than the random forest model, which in turn outperforms the decision tree model.

References

[1] Xuanrui Zhang, Minxia Shi, Cong He, Junhao Li, "On Site Oscillating Lightning Impulse Test and Insulation Diagnose for Power Transformers," IEEE Transactions on Power Delivery, Vol. 3(5), pp. 2548-2550, 2020.

[2] Hanbo Zheng, Enchen Yang, Shuyue Wu, Weijie Lv, Hang Yang, Xufan Li, Xiaoqing Luo, Wei Hu, "Investigation on Formation Mechanisms of Carbon Oxides During Thermal Aging of Cellulosic Insulating Paper," IEEE Transactions on Dielectrics and Electrical Insulation, Vol. 2(4), 2022, pp. 1226-1233.

[3] Shijun Li, Zhao Ge, Ahmed Abu-Siada, Liuqing Yang, Shengtao Li, Kiyoshi Wakimoto, "A New Technique to Estimate the Degree of Polymerization of Insulation Paper Using Multiple Aging Parameters of Transformer Oil," IEEE Access (Volume: 7) pp. 157471-157479.

[4] Hisashi Morooka, Akira Yamagishi, Hideyuki Miyahara, Hiroyuki Sampei, "Investigation into degradation properties of low-viscosity silicone liquid/paper insulation systems for diagnosis in liquid-immersed transformers," IEEE Transactions on Dielectrics and Electrical Insulation (Volume: 24, Issue: 6, December 2017), pp. 3916-3921.

[5] Andrea Cavallini, R. Karthik, Fabrizio Negri, "The effect of magnetite, graphene oxide and silicone oxide nanoparticles on dielectric withstand characteristics of MO," IEEE Transactions on Dielectrics and Electrical Insulation (Volume: 22, Issue: 5, October 2015) pp. 2592-2600.

[6] Zichen Zhang, Khai D. T. Ngo, Guo-Quan Lu, "Characterization of a Nonlinear Resistive Polymer-Nanoparticle Composite Coating for Electric Field Reduction in a Medium-Voltage Power Module," IEEE Transactions on Power Electronics (Volume: 37, Issue: 3, March 2022) pp. 2475-2479.

[7] Anu KumarDas, DayalCh Shill, Saibal Chatterjee, "Coconut oil for utility transformers – Environmental safety and sustainability perspectives," Renewable and Sustainable Energy Reviews Volume 164, August 2022, 112572.
[8] LeylaRaeisian, Hamid Niazmand, Ehsan Ebrahimnia-Bajestan, Peter-Werle, "Feasibility study of waste vegetable oil as an alternative cooling medium in transformers," Applied Thermal Engineering Volume 151, 25 March 2019, Pages 308-317.
[9] T. Mariprasath, M. Murali, S. M. Khaja Moinuddin, S. Khaleef, "An experimental analysis on withstand ability of cellulosic insulating material immersed in NEO for transformer", Materials today proceediongs Volume 52, Part 3, 2022, Pages 1721-1725.
[10] M. Bakrutheen, M. Willjuice Iruthayarajan, A. Narayani, "Statistical failure reliability analysis on edible and non edible natural esters based liquid insulation for the applications in high voltage transformers", IEEE Transactions on Dielectrics and Electrical Insulation vol. 25, pp. 1579-1586, 2018.
[11] T. Mariprasath, V. Kirubakaran, "Feasibility Analysis of Karanja as Alternate Liquid Dielectrics for Distribution Transformers," John Wiley & Sons Ltd, International Transactions on Electrical Energy Systems, Vol. 27, pp. 1-8, DOI: 10.1002/etep.2345, 2017. Impact Factor = 2.860.
[12] Sifeddine Abdi, Ahmed Boubakeur, Abderrahmane Haddad, Noureddine Harid, "Influence of artificial thermal aging on transformer oil properties," Taylor & Francis Electric Power Components and Systems, vol. 39, pp. 1701-1711, 2011.
[13] M. A. Al-Eshaikh, M. I. Qureshi, "Evaluation of food grade corn oil for electrical applications," Taylor & Francis International Journal of Green Energy, vol. 9, pp. 441-455, 2012.
[14] Ruijin Liao, Chao Guo, Ke Wang, Lijun Yang, "Investigation on thermal aging characteristics of vegetable oil-paper insulation with flowing dry air," IEEE Trans. Dielectr. Electr., vol. 20, pp. 1649-1458, 2013.
[15] Jashandeep Singh, Yog Raj Sood, Piush Verma, " The influence of service aging on transformer insulating oil parameters," IEEE Trans. Dielectr. Electr., vol. 19, pp. 421-426, 2012.
[16] M. H. Abderrazzaq, F. Hijaz, "Impact of multi-filtration process on the properties of olive oil as a liquid dielectric," IEEE Trans. Dielectr. Electr. Insul., vol. 19, no. 5, pp. 1673- 1680, 2012.

[17] Q. Liu, Z. D. Wang, "Breakdown and withstand strengths of ester transformer liquids in a quasi-uniform field under impulse voltages," IEEE Trans. Dielectr. Electr. Insul., vol. 20, no. 2, pp. 571-579, 2013.
[18] X. Wang, Z. D. Wang, "Study of dielectric behaviour of ester transformer liquids under ac voltage," IEEE Trans. Dielectr. Electr. Insul., Vol. 19, No. 6, pp. 1916-1925, 2012.
[19] Santanu Singha, Roberto Asano, George Frimpong, C. Clair Claiborne and Don Cherry, "Comparative aging characteristics between a high oleic natural ester dielectric liquid and mineral oil," IEEE Trans. Dielectr. Electr. Insul., Vol. 21, No. 1, pp. 149-158, 2014.
[20] Ruijin Liao, Jian Hao, George Chen, Zhiqin Ma and Lijun Yang, "A comparative study of physicochemical, dielectric and thermal properties of pressboard insulation impregnated with natural ester and mineral oil," IEEE Trans. Dielectr. Electr. Insul., Vol. 18, No. 5, pp. 1626-1637, 2011.
[21] Jung-Il Jeong, Jung-Sik An and Chang-Su Huh, "Accelerated aging effects of mineral and vegetable transformer oils on medium voltage power transformers," IEEE Trans. Dielectr. Electr. Insul., Vol. 19, No. 1, pp. 156-16, 2012.
[22] Daisuke Saruhashi, Xiang Bin, Liu Zhiyuan and Satoru Yanabu, "Thermal degradation phenomena of flame resistance insulating paper and oils," IEEE Trans. Dielectr. Electr. Insul., Vol. 20, No. 1, pp. 122-127, 2013.
[23] Abdul Rajab, Aminuddin Sulaeman, Sudaryatno Sudirham, Suwarno, "A comparison of dielectric properties of palm oil with mineral and synthetic types insulating liquid under temperature variation," ITB J. Eng. Sci., vol. 43, pp. 191-208, no. 3, 2011.
[24] Yu Liu, Jian Li, Zhaotao Zhang, "Fault gases dissolved in vegetable insulating oil under electrical faults," IEEE Conf. on Electr. Ins. Diel. Phen., pp. 198-203, 2013.
[25] Umar Khayam, Abdul Rajab, "Dielectric properties partial discharge properties and DGA of ricinnus oils as biodegradable liquid insulating materials," IEEE Intern. Conf. Cond. Monitor. Diag., pp. 1249-1252, 2012.
[26] Ivanka Höhlein-Atanasova, Rainer Frotscher, "Carbon oxides in the interpretation of DGA in transformers and tap changers," IEEE Electr. Insul. Mag., vol. 26 pp. 22-26, 2010.

[27] N. A. Muhamad, B. T. Phung, "DGA of faults in biodegradable oil transformer insulating systems," IEEE Intern. Conf. Cond. Monitor. Diag., pp. 663-666, 2008.

[28] M. Augusta, G. Martins, A. R. Gomes, "Comparative study of the thermal degradation of synthetic and natural esters and mineral oil: effect of oil type in the thermal degradation of insulating kraft paper," IEEE Electr. Insul. Mag., vol. 28, pp. 22-28, 2012.

[29] N. A. Muhamad, B. T. Phung, "DGA for common transformer faults in soy seed-based oil," IET Electr. Power Appl., vol. 5, pp. 133-142, 2011.

[30] D. Zhou, N. Azis, "Examining acceptable DGA level of in-service transformers," Intern. Conf. on High Voltage Eng. Appl., pp. 612-616, 2012.

[31] Daniel Martin, Nick Lelekakis, "Preliminary results for dissolved gas levels in vegetable oil–filled power transformer," IEEE Electr. Insul. Mag., vol. 26, pp. 41-48, 2010.

[32] X. Wang, Z. D. Wang, J. V. Hinshaw, J. Noakhes, "Comparison of On line and lab dga methods for condition assessment of mineral and vegetable transformer oil," IEEE Conf. Cond. Monitor. Diag., pp. 617-620, 2012.

[33] Zhongdong Wang, Xin Wang, "Gas generation in natural ester and mineral oil under partial discharge and sparking faults," IEEE Electr. Insul. Mag., vol. 29, pp. 62-70, 2013.

[34] Sukhbir Singh, M. N. Bandyopadhyay, "DGA technique for incipient fault diagnosis in power transformers: A bibliographic survey," IEEE Elect. Ins. Mag., vol. 26, pp. 41-46, 2010.

[35] M. Jovalekic, D. Vukovic, "Gassing behaviour of various alternative insulating liquids under thermal and electrical stress," IEEE Intern. Sym. on Elect. Insul., pp. 490-493, 2012.

[36] Imad-U-Khan, Zhongdong Wang, "DGA of alternative fluids for power transformers," IEEE Electr. Insul. Mag., vol. 23, pp. 5-14, 2007.

[37] Satoshi Arazoe, Daisuke Saruhashi, Yuki Sato, Satoru Yanabu, Genyo Ueta and Shigemitsu Okabe, "Electrical characteristics of natural and synthetic insulating fluids," IEEE Trans. Dielectr. Electr. Insul., vol. 18, no. 2, pp. 506-512, 2011.

[38] T. Leibfried, "Online monitors keep transformers in service," IEEE Comp. App. Power, vol. 11, pp. 36-42, 1998.

[39] P. Boss, "Economical aspects and practical experiences of power transformers on-line monitoring," presented at the International Council on Large Electric Systems, France, 2000.
[40] Nick Lelekakis, Wenyu Guo, Daniel Martin, Jaury Wijaya, Dejan Susa, "A field study of aging in paper-oil insulation systems," IEEE Electr. Insul. Mag., vol. 28, pp. 12-19, 2012.
[41] Huo-Ching Sun, Yann-Chang Huang, Chao-Ming Huang, "Fault diagnosis of power transformers using computational intelligence: a review," Energy Procedia, vol. 14, pp. 1226-1231, 2012.
[42] Rui-jin Liao, Chao Tang, Li-jun Yang, Stanislaw Grzybowski, "Thermal aging micro-scale analysis of power transformer pressboard," IEEE Trans. Dielectr. Electr. Insul., vol. 15, no. 5, pp. 1281- 1286, 2008.
[43] Shi-QiangWang, Guan-Jun Zhang, Jian-Lin Wei, Shuang-suo, Yang Ming Dong, Xin-Bo Huang, "Investigation on dielectric response characteristics of thermally aged insulating pressboard in vacuum and oil-impregnated ambient," IEEE Trans. Dielectr. Electr. Insul., vol. 17, no. 6, pp. 1853-1862, 2010,
[44] SanjeebMohanty, Saradindu Ghosh, "Breakdown voltage of solid insulations: its modelling using soft computing techniques and its microscopic study," Elsevier Electrical Power and Energy Systems, vol. 62, pp. 825-835, 2014.
[45] Jiaming Yan, Ruijin Liao, Lijun Yang, Jian Li, Bin Liu, "Product analysis of partial discharge damage to oil-impregnated insulation paper," Elseiver Applied Surface Science, vol. 257, pp. 5863-5870, 2011.
[46] R. Karthik, T. SreeRengaRaja, T. Sudhakar, "Deterioration of solid insulation for thermal degradation of transformer oil," Cent. Eur. J. Eng., Vol. 3, no. 2, pp. 226-232, 2013.
[47] R. Karthik, T. SreeRenga Raja, S. S. Shunmugam, T. Sudhakar, "Performance evaluation of ester oil and mixed insulating fluids," Springer J. Inst. Eng. India Ser. B., vol. 93, pp. 173-178, 2013.
[48] I. L. Hosier, A. Guushaa, E. W. Westenbrink, C. Rogers, A. S. Vaughan, S. G. Swingler, "Aging of biodegradable oils and assessment of their suitability for high voltage applications," IEEE Trans. Dielectr. Electr., Vol. 18, No. 3, pp. 728-738, 2011.
[49] Shivani Hembrom, "Analysis of aged insulating oil for early detection of incipient fault inside the high voltage equipment," Master thesis, National Institute of Technology, Rourkela, May 2013.

[50] Atikah Binti Joharithe, "Effect of long term thermal ageing on the performance of pfae oil as a transformer oil," Bachelor thesis, Faculty of Electrical Engineering, Universiti Teknologi, Malaysia, June 2014.

[51] I. Fofana, A. Bouaïcha, M. Farzaneh, "Characterization of aging transformer oil–pressboard insulation using some modern diagnostic techniques," Euro. Trans. Electr. Power., Vol. 21, No. 1, pp. 1110-1127, 2011.

[52] E. M. Rodriguez-Celis, S. Duchesne, J. Jalbert. M. Ryadi, "Understanding ethanol versus methanol formation from insulating paper in power transformers pyrolysis," Cellulose, vol. 22, pp. 3225-3236, 2015.

[53] A. A. Suleiman, N. A. Muhamad, N. Bashir, N. S. Murad, Y. Z. Arief, "Effect of moisture on breakdown voltage and structure of palm based insulation oils," IEEE Trans. Dielectr. Electr. Insul.., vol. 21, pp. 2119-2126, 2014.

[54] Hasmat Malik, Surinder Singh, Mantosh Kr, R. K. Jarial, "UV/Vis response based fuzzy logic for health assessment of transformer oil," Procedia Engineering, pp. 905-912, 2012.

[55] Norazhar Abu Bakar, A. Abu-Siada, Narottam Das, Syed Islam, "Effect of conducting materials on UV-VIS spectral response characteristics," Journal of Electrical and Electronic Engineering, vol. 1(3), pp. 81-86, 2013.

[56] A. Neffer, Gomez, Rodrigo Abonia, Hector Cadavid, Ines H. Vargas, "Chemical and spectroscopic characterization of a vegetable oil used as dielectric coolant in distribution transformers," J. Braz. Chem. Soc. vol. 22, no. 12.

[57] Abdul Rajab, Aminuddin Sulaeman, Sudaryatno Sudirham, Suwarno, "A comparison of dielectric properties of palm oil with mineral and synthetic types insulating liquid under temperature variation," ITB J. Eng. Sci., vol. 43, pp. 191-208, no. 3, 2011.

[58] Z. H. Shah, Q. A. Tahir, "Dielectric properties of vegetable oils," J. Sci. Res., vol. 3, no. 3, pp. 481-492, 2011.

[59] A. Raymon, P. Samuel Pakianathan, M. P. E. Rajamani, R. Karthik, "Enhancing the critical characteristics of natural esters with antioxidants for power transformer applications," IEEE Trans. Dielectr. Electr., vol. 20, no. 3.

[60] Maik Koch, Stefan Tenbohlen, Tobias Stirlm, "Diagnostic application of moisture equilibrium for power transformer," IEEE Trans. Power Del., vol. 25, no. 4, pp. 2574- 2581, 2010.
[61] Determination of water in insulating liquids and oil impregnated paper, press board by automatic karlfischer titration method- method of test, IS13567. 1992.
[62] Method of test for power factor and dielectric constant of electrical insulating liquids, IS 6262, 2001.
[63] Abdul Rajab, Aminuddin Sulaeman, Sudaryatno Sudirham, Suwarno, "A comparison of dielectric properties of palm oil with mineral and synthetic types insulating liquid under temperature variation," ITB J. Eng. Sci., vol. 43, pp. 191-208, no. 3, 2011.
[64] Method of test for specific resistance (resistivity) of electrical insulating liquids, IS6103, 2006.
[65] Guide for acceptance and maintenance of natural ester fluids in transformers, ieee power and energy society, IEEE C57.147-2008, 2008.
[66] Method of test for petroleum and products, part1, section 1: determination of acid number of petroleum products by potentiometric titration, IS 1448-1-1, 2007.
[67] Method of test for interfacial tension of oil against water by the ring method, IS 6104, 2006.
[68] Petroleum and its products-methods of test, part 21: flash point (closed) by pensky martens apparatus, IS 1448-21, 1992.
[69] T. Mariprasath, V. Kirubakaran, Sreedhar Madichetty, K Amaresh, "An Experimental Study on Spectroscopic Analysis of Alternating Liquid Dielectrics for Transformer," Springer, Electrical Engineering, 2020.
[70] Methods of test for petroleum and its products, Part 16: density, relative density or API gravity of crude petroleum and liquid petroleum products by hydrometer method, IS 1448-16, 2002.
[71] Methods of test for petroleum and its products, part 25: determination of kinematic and dynamic viscosity, IS 1448-25, 1976.
[72] Mineral oil-impregnated electrical equipment in services-guide to the interpretation of dissolved and free gases analysis, IS 10593, 2006.
[73] Hussaian Basha CH, T Mariprasath, Shaik Rafi Kiran, M Murali, "An Experimental Analysis of Degradation of Cellulosic Insulating Material Immersed in Natural Ester Oil for Transformer," ECS Transactions, Vol. 107, pp. 1-8, 2022. DOI 10.1149/10701.18957ecst.

[74] Shaik Rafi Kirana, T. Mariprasath, CH Hussaian Basha, M. Murali, M. Bhaskara Reddy, "Thermal Degrade Analysis of Solid Insulating Materials Immersed in Natural Ester Oil and Mineral Oil by DGA," Materialstoday: Proceedings, August 2021, https://doi.org/10.1016/j.matpr.2021.09.015.

[75] T. Mariprasath, M. Murali, S. M. KhajaMoinuddin, S. Khaleefa, "An experimental analysis on withstand ability of cellulosic insulating material immersed in NEO for transformer," Materialstoday: Proceedings, https://doi.org/10.1016/j.matpr.2021.11.339, 2 December 2021.

[76] Nick Lelekakis, Daniel Martin, "Comparison of dissolved gas-in-oil analysis methods using a dissolved gas-in-oil standard," IEEE Electr. Insul. Mag., vol. 27, no. 5, pp. 29-35, 2011.

[77] T. Mariprasath, V. Kirubakaran, "A Critical Review on the Characteristics of Alternating Liquid Dielectrics and Feasibility Study on Pongamia Pinnata Oil as Liquid Dielectrics," Elsevier, Renewable & Sustainable Energy Reviews, Vol. 65, pp. 784-799, 2016.

[78] T. Mariprasath, V. Kirubakaran, "Thermal Degradation Analysis of Pongamia Pinnata Oil as Alternative Liquid Dielectric for Distribution Transformer," Springer, Sadhana-Academy Proceedings in Engineering Sciences, Vol. 41(9), pp. 933-938, 2016.

[79] V. G. Arakelian, I. Fofana, "Water in oil-filled high-voltage equipment part I: states, solubility and equilibrium in insulating materials," IEEE Electr. Insul. Mag., vol. 27, no. 6, pp. 13-25, 2007.

[80] Ivanka Höhlein-Atanasova, Rainer Frotscher, "Carbon oxides in the interpretation of DGA in transformers and tap changers," IEEE Electr. Insul. Mag., vol. 26 pp. 22-26, 2010.

[81] T. Mariprasath, M. Ravindaran, "An experimental study of partial discharge analysis on environmental friendly insulating oil as alternate insulating material for transformer". Sādhanā- Academy Proceedings in Engineering Sciences, Vol. 47, pp. 1-13, 2022.

[82] T. Mariprasath, V. Kirubakaran, "A Real Time Study on Condition Monitoring of Distribution Transformer Using Thermal Imager," Elsevier, Infrared physics & Technology, Vol. 90, pp. 78-86, 2018.

[83] S. Bagavathiappan, B. B. Lahiri, T. Saravanan, John Philip, T. Jayakumar, "Infrared thermography for condition monitoring – a review," Elseiver Infrared Physics & Technology, vol. 60, pp. 35-55, 2013.

[84] M. Wang, A. J. Vandermaar, K. D. Srivastava, "Review of condition assessment of power transformers in service," IEEE Electr. Insul. Mag., vol. 18, pp. 12-25, 2002.

[85] Sotirios Missas , Xanthi, Greece, Michael G. Danikas, Ioannis Liapis, "Factors affecting the ageing of transformer oil in 150 / 20 kV transformers," IEEE Inter. Conf., pp. 1-4, 2011.

[86] Dhingra Arvind, Singh Khushdeep, Kumar Deepak, "Condition monitoring of power transformer: a review," IEEE Trans. Distr. Conf. Expos., pp. 1-6, 2008.

[87] Esam Al Murawwi, Redy Mardiana, Charles Q. Su, "Effects of terminal connections on sweep frequency response analysis of transformers," IEEE Electr. Insul. Mag., vol. 28, no. 3, pp. 8-13, 2012.

[88] Bhuyan, K. Chatterjee, "Surge modelling of transformer using matlab-simulink," IEEE Conf. Pub., pp. 1-4, 2009.

[89] Xiang Zhang and Ernst Gockenbach, "Determination of the thermal aging factor for life expectancy of 550 kV transformers with a preventive test," IEEE Trans. Dielectr. Electr. Insul., vol. 20, pp. 1984-1991, 2013.

[90] Yunus Biçen, Faruk Aras and Hulya Kirkici, "Lifetime estimation and monitoring of power transformer considering annual load factors," IEEE Trans. Dielectr. Electr. Insul., vol. 21, pp. 1360-1367, 2014.

[91] CIGRÉ Working Group, "An international survey on failures in large power transformers in service," Electra, vol. 88, 1983.

[92] V. I. Kogan, "Failure analysis of EHV transformers," IEEE Trans. Power Del., vol. 3, pp. 672-683, 1988.

[93] Georgia-Ann Klutke, Peter C. Kiessler, M. A. Wortman, "Critical look at the bathtub curve," IEEE Trans. Rel., vol. 52, pp. 125-129, 2003.

[94] O. N. Grechko, I. Kalacheva, "Current trends in the development of in-service monitoring and diagnostic systems for 110-750 kV power transformers," Applied Energy: Russian Journal of Fuel, vol. 34, pp. 84-97, 1996.

[95] Jiaming Yan , Ruijin Liao, Lijun Yang, Jian Li, Bin Liu, "Product analysis of partial discharge damage to oil-impregnated insulation paper," Elseiver Applied Surface Science, vol. 257, pp. 5863-5870, 2011.

[96] Alison K. Lazarevich, "Partial discharge detection and localization in high voltage transformers using an optical acoustic sensor," Master thesis, The faculty of The virginia polytechnic institute and state university, 2003.

[97] Mm Yaacob, Ma Alsaedi, Rashed, Am Dakhil, Sf Atyah, "Review on partial discharge detection techniques related to high voltage power equipment using different sensors," Photonic Sensors., vol. 4, pp. 325-337, 2014.

[98] Csaba Vörös, Gusztáv Csépes, Bálint Németh, István Berta, "Investigation of complex, in-homogenous insulation system with multiple time constant by recovery voltage measurement (RVM)," IEEE Electr. Insul., pp. 35-39, 2013.

[99] A. A. Paithankar, C. T. Pinto, "Transformer insulation diagnosis: recovery voltage measurement and dc absorption test," IEEE Elect. Insul. Electr. Manuf. amp, pp. 597-600, 2000.

[100] Paul Henault, "Detection of internal arcing faults in distribution transformers," Presentation at the ESMO Technical Session, pp. 1-7, 2011.

[101] Zhang Xin, Huang Ronghui, Huang Weizhao, Yao Shenjing, Hou Dan, Zheng Min, "Real-time temperature monitoring system using FBG sensors on an oil-immersed power transformer," High Voltage Engineering, vol. 40, pp. 253-259, 2014.

[102] T. Leibfried, "Online monitors keep transformers in service," IEEE Comp. App. Power, vol. 11, pp. 36-42, 1998.

[103] P. Boss, "Economical aspects and practical experiences of power transformers on-line monitoring," presented at the International Council on Large Electric Systems, France, 2000.

[104] P. Kang, D. Birtwhistle, "On-line condition monitoring of tap changers-field experience," IEEE Conf. Publ., no. 482, pp. 2-6, 2001.

[105] Imad-U-Khan, Z. D. Wang, I. Cotton, S. Northcote, "Dissolved gas analysis of alternative fluids for power transformers," IEEE Electr. Insul. Mag., vol. 23, no. 5, pp. 5-14, 2007.

[106] C. Krause, L. Dreier, A. Fehlmann, J. Cross, "The degree of polymerization of cellulosic insulation: review of measuring technologies and its significance on equipment," IEEE Electr. Insul. Conf., pp. 267-271, 2014.

[107] Daniel Martin, Tapan Saha, David Allan, Kerry Williams, "Extending methods to determine the life remaining of transformer paper insulation," Inter. Conf. Electr. Comp. Eng., pp. 301-304, 2014.

[108] Huo-Ching Sun, Yann-Chang Huang, Chao-Ming Huang, "Fault diagnosis of power transformers using computational intelligence: a review," Energy Procedia, vol. 14, pp. 1226-1231, 2012.

[109] A. S. Nazmul Huda, SoibTaib, "Application of infrared thermography for predictive/ preventive maintenance of thermal defect in electrical equipment," Applied Thermal Engineering, vol. 61, pp. 220-227, 2013.

[110] Zhikai Xing, Yigang He, Jianfei Chen, Xiao Wang, Bolun Du, "Health evaluation of power transformer using deep learning neural network," Electric Power Systems Research, Vol. 215, Part B, 109016, 2023.

[111] Dhanu Rediansyah, Rahman Azis Prasojo, Suwarno, A. Abu-Siada, "Artificial Intelligence-Based Power Transformer Health Index for Handling Data Uncertainty," IEEE Access, Vol. 9, 2021.

[112] Han, Yu, and Y. H. Song. "Condition monitoring techniques for electrical equipment-a literature survey." IEEE Transactions on Power delivery 18.1 (2003): 4-13.

[113] Wani, S. A., Rana, A. S., Sohail, S., Rahman, O., Parveen, S. and Khan, S. A., 2021. Advances in DGA based condition monitoring of transformers: A review. Renewable and Sustainable Energy Reviews, 149, p. 111347.

[114] Rediansyah, D., Prasojo, R. A., & Abu-Siada, A. (2021). Artificial intelligence-based power transformer health index for handling data uncertainty. IEEE Access, 9, 150637-150648.

[115] Esmaeili Nezhad, A., & Samimi, M. H. (2022). A review of the applications of machine learning in the condition monitoring of transformers. Energy Systems, 1-31.

[116] Ganyun, L. V., Haozhong, C., Haibao, Z., & Lixin, D. (2005). Fault diagnosis of power transformer based on multi-layer SVM classifier. Electric power systems research, 74(1), 1-7.

[117] Hua, Y., Sun, Y., Xu, G., Sun, S., Wang, E., & Pang, Y. (2022). A fault diagnostic method for oil-immersed transformer based on multiple probabilistic output algorithms and improved DS evidence theory. International Journal of Electrical Power & Energy Systems, 137, 107828.

[118] Hu, Wei, et al. "Loss reduction strategy and evaluation system based on reasonable line loss interval of transformer area." Applied energy 306 (2022): 118123.

[119] Asghar, Rafiq, et al. "Reduction of distribution system losses through WAPDA distribution system line-loss reduction program." Mehran University Research Journal Of Engineering & Technology 41.2 (2022): 79-90.

[120] Gao, Meng, Mingli Fu, Ran Zhuo, Lei Jia, Yuan La, and Lianhong Zhong. "Carbon Footprint of Power Transformer by Life Cycle

Assessment." In 2022 IEEE International Conference on High Voltage Engineering and Applications (ICHVE), pp. 1-4. IEEE, 2022.

[121] Shengwei, C. A. I., Y. I. N. G. Si, W. A. N. G. Xinsheng, W. E. I. Zhixiong, N. I. U. Yanzhao, Z. H. A. N. G. Qian, and G. O. N. G. Yujia. "Calculation and analysis of life cycle carbon emissions of distribution transformers." In 2022 IEEE International Conference on High Voltage Engineering and Applications (ICHVE), pp. 1-4. IEEE, 2022.

[122] Serres, Hugo. "Life Cycle Assessment of typical projects of the distribution power network: Assessment, Improvement & Recommendations." (2022).

[123] Laayati, O., El Hadraoui, H., El Magharaoui, A., El-Bazi, N., Bouzi, M., Chebak, A., & Guerrero, J. M. (2022). An AI-Layered with Multi-Agent Systems Architecture for Prognostics Health Management of Smart Transformers: A Novel Approach for Smart Grid-Ready Energy Management Systems. Energies, 15(19), 7217.

[124] Youssef, M. M., Ibrahim, R. A., Desouki, H. and Moustafa, M. M. Z., 2022, March. An Overview on Condition Monitoring & Health Assessment Techniques for Distribution Transformers. In 2022 6th International Conference on Green Energy and Applications (ICGEA) (pp. 187-192). IEEE.

[125] Esmaeili Nezhad, A. and Samimi, M. H., 2022. A review of the applications of machine learning in the condition monitoring of transformers. Energy Systems, pp. 1-31.

[126] Alabdullh, M. K. K., M. Joorabian, S. G. Seifossadat, and M. Saniei. "A New Model for Predicting the Remaining Lifetime ofTransformer Based on Data Obtained Using Machine Learning." Journal of Operation and Automation in Power Engineering (2023).

[127] Xu, D., Xiao, X., Liu, J., & Sui, S. (2023). Spatio-temporal degradation modeling and remaining useful life prediction under multiple operating conditions based on attention mechanism and deep learning. Reliability Engineering & System Safety, 229, 108886.

[128] Singh, Harkamal Deep, and Jashandeep Singh. "Experimental Analysis on Aging Evaluation of Transformer Oil via Data Collection." In Modern Electronics Devices and Communication Systems: Select Proceedings of MEDCOM 2021, pp. 383-395. Singapore: Springer Nature Singapore, 2023.

Index

About the Authors

Dr. T. Mariprasath received his Ph.D. degree from the Rural Energy Centre, The Gandhigram Rural Institute (deemed to be a university) in January 2017, fully funded by the Ministry of Human Resource Development âĂŞ Government of India. He has been working as an Associate Professor with the Department of EEE, K.S.R.M. College of Engineering (Autonomous), Kadapa since June 2018. He has published in 10 science-citation-indexed journals and 15 Scopus-indexed articles. He had an Indian patent grant and an Australian innovation patent grant. He received Rs 6 lakh from MSME, Government of India, to develop a self-powered GPS tracker. His research interests include renewable energy resources, power electronics, green dielectrics, and artificial intelligence.

Dr. V. Kirubakaran is an Associate Professor in the Rural Energy Centre at the Gandhigram Rural Institute, which is deemed to be a university in India. He obtained a Ph.D. in energy engineering from the National Institute of Technology, India, in May 2007. He is a recipient of the Young Scientist Fellowship, awarded by the Department of Science and Technology of the Government of India. He has published more than 75 papers in various national and international journals, in addition to authoring seven books. He had two Indian patent grants. His areas of interest include energy engineering and gasification.

P. Saraswathi is working as an assistant professor in the department of physics at A.P.C. Mahalaxmi College for Women, India. She completed a master of philosophy in physics in 2013 at M.S. University Tirunelveli. Also, she completed a Master of Science in physics at Pope Arts and Science College in 2011. She had an Indian patent grant in the field of biodegradable materials. She has materials science, atomic and nuclear physics

Dr. Kumar Reddy Cheepati received his Ph.D. degree from JNTUK, Kakinada, India, in 2021. He is currently working as an associate professor with the Department of Electrical and Electronics Engineering, KSRM College of Engineering, Kadapa, Andhra Pradesh, India. He has 11 years of academic experience. He has published research papers in various international journals of high repute, which include Scopus, SCI, and ESCI-indexed journals. Also, he published an Indian patent. His area of research includes power electronics, electrical vehicles, and power quality.

Dr. Prakasha Kunkanadu Rajappa received his Ph.D. degree from CSIR-Central Electrochemical Research Institute-Chennai Unit. He is currently working as a scientist in Catalonia Institute for Energy Research IREC, Barcelona, Spain. He had an Indian patent grant in the field of materials science and has also published more than 10 patents. In addition, he has published 20 research articles in reputed journals. His area of research includes material science, and battery and electrical vehicle technology

For Product Safety Concerns and Information please contact our EU representative GPSR@taylorandfrancis.com Taylor & Francis Verlag GmbH, Kaufingerstraße 24, 80331 München, Germany